Getting In

The Insider's Guide to Finding the Perfect Undergraduate Research Experience

David G. Oppenheimer, PhD

Paris H. Grey

Department of Biology, UF Genetics Institute, and
Plant Molecular & Cellular Biology Program
University of Florida
Gainesville, FL

Secret Handshake Press
2015

Copyright © 2015 David G. Oppenheimer and Paris H. Grey

All rights reserved. No part of this publication may be reproduced, distributed, or transmitted in any form or by any means, including photocopying, recording, or other electronic or mechanical methods, without the prior written permission of the publisher, except in the case of brief quotations embodied in critical reviews and certain other noncommercial uses permitted by copyright law. For permission requests, email to the publisher, addressed "Attention: Permissions Coordinator," at the address below.

Published by Secret Handshake Press
10311 NW 13th Avenue
Gainesville, FL 32606

secrethandshakepress@gmail.com

Library of Congress Control Number: 2015947336
ISBN-13: 978-0692488348
ISBN-10: 0692488340

First Edition

In memory of Dr. D. Peter Snustad

Acknowledgments

We are especially grateful to our former undergraduate, Madeleine Turcotte, for discussions, unvarnished opinions, and three years of unwavering dedication to her research projects, even when faced with challenges, and when there was always one more protein gel to pour and run. We wish her the best in her adventures at the University of Oxford, in medical school, and in everything she pursues after.

Special thanks to Dr. Dan Purich for enthusiasm, perspectives, an insider's knowledge about publishing, and for the two most important words in the title.

Our deepest gratitude to Linda Howard, a dear friend to us both, who everyday shows us the importance of sharing both knowledge and compassion to make the world a better place.

We thank countless others who shared their experiences about being an undergraduate researcher (or working with an undergraduate in the lab), and of course thanks to all the students we have mentored, trained, or supervised during our careers so far.

David G. Oppenheimer, PhD
Paris H. Grey

I wish to thank my parents, who always offer encouragement.

D.G.O.

I wish to thank my friend and colleague, Dr. Brandi Ormerod, for being awesome, and for her enthusiasm, suggestions, and that key bit of insight when I needed it the most. To my godmother, Susan Tan, for always being supportive and having an encouraging word ready to go whenever I need it.

P.H.G.

Contents

Introduction

Depending on your major and career path, you may have already heard numerous times that it's important to participate in undergraduate research, especially if you are a student in science, technology, engineering, or mathematics (STEM). But if an undergraduate research experience is so important (and it is), and has so many potential benefits (which it does), why is it so difficult to find a lab and get a research position?

The simple fact is that undergraduate research positions are highly competitive, and that is unlikely to change. Many medical and graduate admissions committees place a high value on research experience, so more students than ever are adding it to their curriculum. This means that the student who has insider information will have a competitive edge when searching for a research position.

Getting In is your competitive edge. From the first contact with a potential research mentor and through the interview process, we teach you what you need to do, tell you what mistakes to avoid, and show you how to be your most professional self.

But we wrote *Getting In* to be much more than a standard how-to guide. As academic insiders, we know that a successful search is about more than contacting professors to ask for a research project. That's why we include information on getting organized, understanding what research is, and evaluating an available position in relation to your goals. We also include sections on the benefits of undergraduate research, an introduction to lab culture, and address pressing questions such as these: "When should I start my search?" and "How much experience do I need?" and "What if my GPA isn't as high as those of my classmates?" and "What if I make a mistake?"

Whether you want to participate in an undergraduate research experience to help you at the next level, because you like the idea of research, or aren't sure if it is even right for you, *Getting In* will help you be a more competitive applicant and help you choose the position that will give you the most meaningful use of your time.

As an associate professor with an active research lab, and a practicing research scientist, we each bring a different perspective to *Getting In*–one from the professor's desk, and one from the researcher's bench. We've combined our perspectives to make a comprehensive guide so whether you interview with a professor or a researcher, you'll be prepared.

You'll also notice that throughout *Getting In,* we interchange the pronouns *she, he,* and *they* to represent a singular person, because there is still no pronoun in the English language that represents us all. Although many grammarians will no doubt disagree with our decision, we made this stylistic choice to be inclusive and hope you embrace it as so.

We wish you the best of luck with your search for a research experience.

Now, let's get started!

CHAPTER 1

A Bold Strategy

The Boldface Text Will Save You Time

YOU'RE busy.

As a student, you strive to create a balance among your academics, personal time, and extracurricular activities, and now you're planning to add a research project into the mix. *Getting In* was written with this balancing act in mind.

You'll notice an abundance of bold text in *Getting In*—probably more than you've seen in any other book. The bold text is what you would highlight if this was a textbook. It's the information I (DGO) emphasize when teaching undergraduates about how to find the perfect research experience. **Although all the information in *Getting In* is important, the bold was strategically selected to help you learn the essential information quickly.**

Therefore, if you're short on time, or need to quickly review a section, focus on the bold for the essential information and read the rest to understand *why* the bold is essential, for examples, or how a strategy will help you succeed.

CHAPTER 2

Why Choose Research?

What's in It for Me?

IF you're considering undergraduate research, you probably already know some of the potential benefits: looks good on your resume, can help you explore a potential career path, and can lead to recommendation letters. **However, an undergraduate research experience can also support your long-term professional goals, regardless of what your career path turns out to be.**

For example, if your career goals include graduate school, pursuing an MD–PhD, professional school, or research during a medical residency, an undergraduate research experience will help you prepare. In addition to acquiring basic (and possibly advanced) research skills, you'll gain insight from being part of a professional lab environment.

Furthermore, admissions committees value applicants who have participated in inquiry-based activities and have critical thinking skills, intellectual curiosity, creativity, initiative, and the ability to produce results. An in-depth research experience will provide ample opportunities for you to develop and demonstrate these qualities to potential references. If you're headed to graduate school, you'll gain an additional competitive advantage because many admissions committees believe that success in undergraduate research is a good indicator of future success in graduate school. As a bonus, an in-depth undergraduate research experience can help your application stand out when applying for graduate fellowships.

If your career goals include entering the research job market directly after your undergraduate degree is awarded, obviously research skills and experience working in a lab will give you a competitive edge. However, even if you ultimately choose a non-research career path, the qualities you develop during your research experience will give you a career advantage. For example, in the lab you'll likely have countless opportunities to demonstrate that you're reliable, work well with others, and are able to contribute to team and project goals. Virtually every employer wants to hire the candidate with a proven track record in those areas. And if your long-term goals include an entrepreneurial track, owning your own clinic, or directing your own lab, those skills will be essential for your success.

Even with all the benefits just stated, you might be surprised to learn that what you gain from a meaningful, in-depth research experience could be much more. **A research experience can foster personal development, give you an academic edge, create connections, and potentially lead to a fellowship or internship.** In addition, you might find that learning to apply scientific principles at the research bench teaches you how to apply an analytical approach to solving personal problems, evaluating options when presented with a difficult decision, or examining the world at large.

How to Get the Most Out of Your Research Experience

To get the most out of your research experience, you'll need to do more than show up and put in the hours. You'll need to make the effort to learn about what's going on around you, and you'll need to embrace that your biggest intellectual accomplishment might not come from understanding a result, but from thinking of the next question to ask. In other words, to get the most out of your research experience, you'll need to invest yourself in the commitment.

How will you do that? **Make your research experience a priority. Show up on time, ready to work and contribute. Do a little more than the required minimum each day and adopt "What more can I learn?" as a personal philosophy.**

If you choose this approach to your research experience, you'll gain the most benefits and be rewarded with the most fulfilling and meaningful use of your time.

Professional Development Opportunities

Much of professional development can be classified into one of two categories: (1) accomplishments listed on a CV (curriculum vitae) or resume, or (2) skills that admissions committees and potential employers highly value. Items from either category can be used in letters of reference or recommendation, but items in the second category have the additional bonus of contributing to your success outside the lab, in college, and in your chosen career path, regardless of what it turns out to be.

Items from Category 1—Accomplishments That Can Be Listed on a CV or Resume

Here are three examples of research experiences with varying levels of time commitment and the potential professional development to be gained from each. The benefits you gain will vary depending on a variety of factors such as your project, your training opportunities, and your investment in your research experience. **As you read through each experience, highlight items you'd ideally like to be able to include on your resume or CV at the end of your research experience.** (You'll use this information later.)

Research experience example 1

- Three to six hours per week
- One or two semesters (or less time)
- Lab experience not necessarily continuous (For example, you might participate in Lab A in the spring semester and in Lab B during the summer or fall semester.)

If you participate in a short-term research experience for a few hours a week, and have minimal responsibilities, you might secure a letter of reference and be able to list some of the following accomplishments on your CV or resume:

- Familiarity with many laboratory techniques or expertise in a few
- Specialized knowledge
- Attendance at an undergraduate research conference or symposium, such as one sponsored by a student group, department, or other campus program
- Presentation (poster or a short talk) at an undergraduate research symposium
- Participation in a research experience

Research experience example 2

- Nine to twelve hours per week during the semesters

- Two to four semesters — ideally continuous and in the same lab
- May include a full-time summer research experience in your primary lab or elsewhere

OR

- One or more full-time summer research experiences in the same lab or in multiple labs (possibly as part of a fellowship or internship program)

An increase in lab hours and advanced commitment to your research project will result in additional accomplishments to add to your CV or resume and will likely boost a letter of reference to a strong letter of recommendation. Additional accomplishments may include

- Technical expertise and skills acquired in laboratory techniques
- Leadership skills from training other undergraduates
- Experience in scientific software packages such as those used for clone design, DNA sequence manipulation, and statistical analysis
- Leadership skills and campus outreach such as becoming an ambassador for undergraduate research on your campus through your department or college or as a peer advisor
- Participation in an undergraduate research conference, meeting or symposium such as those sponsored by student groups, departments, or other campus programs
- Participation or presentation at a national undergraduate research conference
- Internship or paid research position at a company or government institution, other university, or grant funded in your primary lab
- Student publication such as a senior thesis or a paper in an undergraduate research journal
- Fellowships, scholarships, or awards for research, travel, or non-science awards

Research experience example 3

- Twelve to fifteen-plus hours per week during the semesters
- One and a half to two-plus years — ideally continuous and in the same lab
- Includes *at least one* full-time summer research experience — ideally in the same lab as your regular semester lab experience

OR

- Several full-time summer research experiences (even without research during the semester, and not necessarily in the same lab—possibly through a fellowship or internship program)
- Full-time research experience for a semester or year that allows a student to focus exclusively on research without taking classes, typically as part of a fellowship program

If you excel and participate in a long-term and in-depth research experience, you'll likely have most of the previously mentioned accomplishments and skills to list on your CV or resume, possibly authorship on a peer-reviewed journal publication, and an outstanding (or epic!) letter of recommendation. In addition, you may also be able to list

- Technical expertise with equipment
- Presentation (poster or an undergraduate talk) at a national or international meeting
- Participation (and presentation) at national or international undergraduate research conferences
- Advanced responsibilities in the lab

Items from Category 2—Skills That Pay the Bills

Here are some of the professional skills and qualities that you may develop, refine, and demonstrate during your undergraduate research experience. How much you gain will depend on your commitment level, personal investment, and the opportunities available through your specific research experience.

Upholding a commitment

Millennials have acquired the reputation of being quick to move between positions and drop activities that are difficult rather than invest their talents and skills in a long-term commitment. **Long-term involvement in an undergraduate research experience will unequivocally demonstrate that you can stick with something** even through the rough spots (which all research projects inevitably have).

Working well with others

Whether it's with your peers, the professional research staff, or your research mentor, your effectiveness in working with others is essential in the lab and critical to your success outside of college. Admissions committees and employers know that misunderstandings stemming from an intolerance or insensitivity to personality or cultural differences can wreak havoc on morale and productivity within a group. **Those who have demonstrated the ability to coexist with all group members, respect the opinions of others, and engage in collaborative efforts are highly valued.** This can be harder to do than you'd think, because chances are you won't like everyone you work with in the lab (or elsewhere for that matter).

Producing results

Of course you know that effort is important for success; however, the results you produce are often the measure used to determine what value those efforts bring to a team, group, or project. Your research experience will present ample opportunities to either take the time to produce quality results or rush through a process just to get it done. If you consistently choose the former, you'll contribute results others rely on and develop the patience and self-discipline to help succeed in your career.

Embracing feedback

Everyone wants to work with someone who wants to learn. An undergraduate who responds to feedback from her research mentor with a positive attitude and the determination to improve will have more help, research opportunities, and success than an undergraduate who responds with defensiveness. **Use your undergraduate research experience to strengthen your ability to listen to feedback, even if it is negative.** This practice will better prepare you for the future when feedback from a supervisor, and your reaction to it, will have the power to affect your career.

Learning research skills

Everyone acquires new knowledge in college, but not everyone learns new skills. Through continued involvement in, and success with, your research project, you'll acquire a specialized skill set. Every admissions committee and employer values candidates who have both the desire and proven ability to learn new skills.

Demonstrating leadership

If you're asked to help train, direct, or supervise other undergraduates, the line on your CV might be "trained # people in technique X," but admissions committees and potential employers know that it takes more than knowledge or proficiency to be an effective instructor — it takes leadership. **Having the responsibility of teaching a technique is good. Using that opportunity to refine and demonstrate your leadership skills is even better.** The more practice you have with this in college, the easier it will be to step up to leadership opportunities after your undergraduate work and in the beginning of your career.

Following instructions

It sounds easy, right? However, many new undergraduates try to "improve" a protocol by changing or skipping a step—sometimes without even realizing it. This usually results in a failed experiment and frustration. **Undergraduates who excel at following protocols make the most progress at the research bench and spend less time redoing experiments due to operator error.** They are also more likely to be given advanced responsibilities or the option of pursuing an independent research project. In every career path or professional position, you'll need to follow instructions. If you polish this ability in the lab, you'll have the skills to go further in any career you choose.

Focusing in a chaotic environment

Lab environments can be loud, chaotic, or distracting, and it can be difficult to focus on the simplest task—especially if labmates' conversations are more interesting. **Learning to block out distractions is an essential skill to master, and it takes a lot of practice to do it well.** In many careers, you won't have a distraction-free environment to perform complicated tasks that are important to you (or your supervisor or a patient), and you won't have the option of putting earphones in to help you focus. Use your research experience to refine your concentration skills in a chaotic environment, and you'll be better off in all professional endeavors.

Mastering critical thinking and problem solving

By taking ownership of your project, whether through optimizing a new technique, troubleshooting a failed experiment, interpreting your results, planning the next step of an experiment, or performing data analysis, **you'll develop the ability to think critically, evaluate problems, and seek solutions**. When a technique doesn't work, you'll learn to evaluate what likely went wrong and use your knowledge to determine what to do about it. You'll learn to narrow the possibilities, determine which option is the best approach, and make a strategy to move forward. You'll also learn to evaluate if your strategy was successful, or if a result leads you to a different conclusion and the need to start again. These skills are not only highly valued by admissions committees and employers, but also will be valuable throughout your life.

Developing organizational strategies

In a nutshell, the better organized you are, the more successful you'll be. **In both your research notebook and wet or dry benchwork, you'll have nearly infinite opportunities to establish, refine, and demonstrate**

organizational skills. No matter the career path you choose, success will come easier and faster if you build solid organizational strategies in college. Use your research experience to create the foundation that will help you succeed.

Planning

Whether you take a complicated protocol and break it down into small, manageable steps, or anticipate future needs for an experiment, **undergraduate research will present myriad opportunities to refine your ability to plan ahead.** There is no way to overestimate the advantage of being able to imagine yourself in a future situation and devise a plan to cover the most likely contingencies, no matter your career path.

Managing time

Everything from arriving on time and keeping a consistent schedule, to knowing how to allocate your time in the lab will help you refine your time-management skills. In every career and profession, excellent time-management skills are an asset. In many professions the lack of such skills will definitely hold you back.

Demonstrating a strong work ethic

Your work ethic reflects not only how hard you work, but also the value you place on the work you do. Being punctual, reliable, and determined to complete the responsibilities of your research project will speak more about a solid work ethic than any guarantee you'll ever make on an application or write about in a personal statement. There is no job or career where a solid work ethic is a disadvantage. Whether your research project hits a roadblock, or enters a stage so boring it's hard to stay self-disciplined, **the work ethic you choose to develop in college (and research in particular) can help push you toward success, or hold you, and your career, back.**

Developing outstanding communication skills

Whether through Q&A sessions with your research mentor, presenting a research poster, or explaining your latest result at a lab meeting, the sooner you start a research experience, the more opportunities you'll have to refine your communication skills. Even if you're one of the lucky few who feel comfortable speaking to a group of strangers, you'll still need your audience to care about your message, or it won't be meaningful. **Because there is no shortcut to developing outstanding communication skills, you should use your undergraduate research experience to help learn to do the following:**

Tailor your message to your audience. How you discuss your research project will vary depending on who you are speaking with and their background knowledge. For example, your research mentor will have advanced knowledge about your project, so you'll use technical language and known abbreviations when you summarize a result or discuss a strategy. However, if you discuss your research with a classmate or present it at a symposium, you'll need to choose a less technical approach to get your message across for the conversations to be valuable. **The ability to adapt your message to your audience, without being condescending, is an essential skill, whether you become a professor, doctor, executive, or entrepreneur.** It's a critical skill to master especially if you choose a career where you will interact with those who have less knowledge about a subject than you do.

Communicate under pressure. At the start of your research experience, conversations with your research mentor will be nerve racking at best. Whether you're asked to interpret a result, explain a mistake, or give your opinion on the next experimental step, it will be easy to panic and difficult to find the right words to form an answer. **Learning to steady yourself and give a clear and concise answer will initially be a challenge, but with enough practice, it will become easy.** Really. You want as much experience with this as possible before you're an intern being grilled by the attending physician, an assistant professor teaching your first class, or an employee explaining to your boss why you deserve the promotion over a colleague who has been at the company longer. Take full advantage of the opportunities during your research experience to polish this skill.

Refine listening and note-taking skills. **The ability to process spoken information and recreate it accurately in your own words is a skill that you'll need in every profession, but one that only comes with practice.** Currently, if your core note-taking technique is to supplement your professors' lecture slides, you could graduate from college without refining this skill. In the lab, you'll need to take notes while your research mentor explains the next step in an experiment, the project objectives, or a change in the protocol—without the benefit of preprinted slides. In addition, when you write up your notebook, create a poster, or write a research report, you'll need to include enough details to make an impact and be useful but not so many that your message gets lost. The longer you're in a research experience, the more you'll refine these communication skills.

Prepare for *the* interview. To take full advantage of your research experience, you'll want to present your research every chance you get. **Each**

time you present your research poster at a symposium or conference, it will seem like an interview (perhaps *interrogation* is the more accurate word). **These "practice interviews" will help you prepare for those oh-so-important interviews near the end of your college career.** If you present your research frequently, by the time you head to your mock interview for med, grad, or professional school, your interview style should only need a little polishing.

Polishing the professional you

Everything listed previously is relevant to this point. Every opportunity you have to develop, refine, or demonstrate your professional growth is a professional advantage. For many undergraduates, a research lab is their first professional work environment and an important opportunity to navigate the expectations of a supervisor while thriving in a work culture. **A research experience is the perfect opportunity to gain the skills you'll need to be successful in the next stage of your career and polish your professionalism.**

Getting a job

Depending on the skills you acquire during your undergraduate research experience, **you may be employable as a research scientist after graduating with a bachelor of science degree.** Your options might include employment at a university research lab, for a company, in industry, at a hospital, in a government lab, or as a researcher in a postbaccalaureate program. Several of our former undergraduates have been employed in professional research positions after graduating with a BS degree—some during a gap or personal year, and others as a stepping-stone in their research career.

Personal Development Opportunities

Some personal and professional development characteristics overlap. For example, self-motivation is an advantage in every professional pursuit; however, it is equally important to achieve personal goals and to prioritize the activities that matter to you the most. It's also important to note that many personal development characteristics support the development of others. For instance, mastering self-motivation makes it easier to stay self-disciplined, and practicing self-reliance makes it much easier to master decision-making.

Personal development happens over time through new experiences and challenges within those experiences. As you develop your intra- and interpersonal skills, you'll learn to rely on your intellect, abilities, and strengths to solve problems and manage your life.

Participation in a research experience doesn't automatically guarantee personal development, but it does give you continuous opportunities to challenge yourself. It's up to you to decide how to use those opportunities.

Personal Fulfillment

Your research experience should be a meaningful and rewarding use of your time. Although there are several ways this can be accomplished, here are some examples.

Making a difference or contributing to a common goal

For many undergraduates, their research experience is an opportunity to find personal fulfillment by participating in something bigger than themselves. Whether researching a disease, understanding an important pathway, or studying the regulation of a specific protein, **knowing that your contributions could have an impact beyond your personal gain is unequaled.**

Defining a career path

A research experience can help you clarify your career path. Is a career in research right for you? Are you trying to decide among medical school, graduate school, or another professional school? Are you in the right discipline? Have you declared the right major? Maybe research on a microbiology project will inspire you in a way that your major never has. Perhaps after spending time in the lab "just to see," you'll consider pursuing a graduate degree or MD–PhD. **Participating in research can help you determine if you're on the right career path, or if you should consider taking a detour.** Even if you do not discover something groundbreaking at the research bench, you'll likely discover how interested you are in a career in science.

Boosting self-confidence

In every area of personal and professional development, there is a potentially hidden bonus—the increase in self-confidence that comes from earned accomplishments. *Your* accomplishments. Whether by gaining technical skills, understanding your project and explaining it to your peers, or solving complex problems at the research bench, you can derive a great deal of satisfaction from accomplishing something that wasn't easy.

Self-Management Skills

Self-management is the ability to be persistent in planning your life, and to follow through with activities that are important to you—especially when faced with challenges and frustration. **Developing self-management skills will help you identify and pursue your passion and build the life you want.** Here are some of the self-management skills you can gain from an undergraduate research experience.

Understanding self-motivation

Self-motivation is your reason or motive for participating in anything—a road trip with friends, a study group, an extracurricular activity, or a research experience. **Understanding your self-motivation is an essential step in taking control of your life and your future.** Perhaps your self-motivation for participating in research comes about because you're inspired by scientific pursuits, and you want to cultivate professional and personal development, earn an epic letter of recommendation, or make your application to medical or graduate school as competitive as possible. The key to self-motivation is making sure your reasons are truly about your desires, your wishes, and your goals. **It's much more difficult to achieve goals if you don't have a clear understanding of why you should pursue them, or if they are actually someone else's goals.** Once you learn to evaluate (and occasionally reevaluate) your self-motivation for undergraduate research, it will help you apply the same accountability to all of your activities.

Strengthening self-discipline

Self-discipline is the ability to pursue your goals even when it's difficult, frustrating, or boring, or when it would be easier to just give up. When you hit a rough spot with research, and your self-motivation isn't enough to get you through it, self-discipline will help you get the work done regardless of how uninspired you feel. Self-discipline is essentially adopting the philosophy, "I'm going to keep working toward my goals and stay focused, no matter what." It is the ability to find the resolve to keep pushing forward even when you really, really, really don't want to—not because you'll "get in trouble" if you don't, but because you have the perseverance to accomplish your goals. **If you use your research experience to strengthen self-discipline while in college, you'll be less likely to give up on your dreams when your life hits a rough spot.**

Improving self-reliance

Self-reliance is the ability to use your personal resourcefulness to complete your responsibilities and accomplish tasks. In a research position, this covers everything from learning where reagents are stored, to knowing what do to when you arrive at the lab each day, and ultimately designing your own experiments. **Self-reliance involves making a conscious decision to try to solve a problem or answer a question instead of instinctively asking someone else for help.** You can be self-reliant and still receive assistance from others, as the two are not mutually exclusive. The key is to balance what you can do for yourself with requests for help from others. **It takes time and practice to be able to do this well.** A research experience can guide you toward self-reliance, but it will be up to you to embrace the challenge to improve it. The more you practice self-reliance, the easier it will become and the more it will be a part of your approach to life.

Developing strategies to cope with failure and disappointment

The one guarantee about research is that sometimes things fail. It could be a technique, an experiment, or a straightforward should-have-been-easy-to-do dilution. **Occasional frustration and disappointment at the bench is a reality for most researchers, and developing strategies to cope with it will be essential to your success.** If you're able to do this consistently throughout your research experience, you'll develop and reinforce positive strategies to use in your life outside the lab. **Those who are resilient when faced with failure and disappointment live happier lives overall, because they know that both are temporary and cannot destroy true determination.**

Taking Risks to Overcome Fear

Because most new experiences involve getting out of one's comfort zone, most undergraduates experience a "healthy fear" at the start of a research experience. This fear, when properly managed, can lead to personal development and lessons learned about oneself along the way. **The most common fears undergraduates conquer through involvement in an in-depth research experience are these:**

The fear of starting something new

Especially at the beginning of a research experience, the lab is an intimidating place—even when everyone is nice. You might be afraid that everyone is judging you on a personal level (they aren't). You might be afraid of interpreting your results for your research mentor (it will get easier with practice). You might be afraid that you'll never become self-reliant like the other undergraduates are (with hard work and a willingness to learn, you will). **Through patience and determination, you'll learn to push through your initial nervousness and begin to feel like a member of the lab.** Then the next time you join a club, start a volunteer experience, or begin an extracurricular activity, it will be easier because you'll already have experience transitioning to a new environment at the college level.

The fear of making the wrong decision

Determining your options for troubleshooting a technique or designing an experiment is only the first step to accomplishing your research objectives. Making the decision of which option to choose can be much harder. By understanding your project and relying on your knowledge, skills, and abilities, you'll learn to consider the "what ifs" to create strategies, make decisions, and commit to an action plan. **Overcoming the fear of making the wrong decision won't be easy, and it won't happen all at once, but through a commitment to your research experience, it will happen.** Once it does, you'll find that the habit sticks with you outside the lab as well.

The fear of making a mistake

Mastering the fear of decision-making and the fear of making a mistake go hand in hand. For most new lab undergraduates, after making a research decision, the greatest source of stress is the fear of making a mistake. We assure you, it will happen. However, when it does, **you'll learn firsthand that making a mistake can't destroy you, and it doesn't mean that you are less smart or capable than you thought.** Once you fully experience this in the lab, it will help you put all mistakes in perspective—both in and out of the lab. That perspective is the key to a less-stressful and happier life.

The fear of looking "stupid"

At the start of a research experience, it might be difficult to ask for help with a protocol or to interpret a result for your research mentor. **Each time you don't let the fear of bringing a mistake to someone's attention, or being incorrect prevent you from answering a question, you'll take a step toward mastering that fear.** The more often you put yourself "out there" during your research experience, the more your self-confidence will grow, and the less stupid you'll feel. As a bonus, your participation in research will make it easier to interact with professors in office hours and ask the questions you need answers to during class.

The fear of change

At times it will seem as if your research experience is an endless string of spontaneous events that you can't control: Your cultures didn't grow. Your plasmid didn't cut. Your cell lines got contaminated. A new result changes the direction of your project. Your research mentor randomly instructs you to process samples with a protocol you've never done before, and he won't be around to help. It can be disconcerting not to have a defined plan of what will happen each day in lab or to need to change research strategies in the middle of an experiment. **If you can learn to embrace change in the lab, instead of being frustrated or overwhelmed, you'll develop strong personal adaptability.** This will make you a happier person overall in the lab, out of the lab, and in life.

Developing an Academic/Life Balance

To get the most out of your college experience, you'll need to develop the right academic/life balance. Here are several ways your undergraduate research experience can help you achieve this commonly sought-after goal.

Using organizational strategies learned in the lab as a model for personal success

It's not easy to get and stay organized—that's why there is an entire industry dedicated to tips and tricks on how to organize one's life and career. For many undergraduates, organizational skills develop slowly, through a process of trial and error, often punctuated by stressful moments caused by overcommitment. Although you won't do protein analysis or complete an actin polymerization assay in your life outside of research, **if you choose to apply the same organizational strategies that help you get experiments done in the lab to your life pursuits, you'll get further ahead with less stress** than someone who doesn't have the advantage of a hands-on research experience to emulate.

Refining time-management skills

Having enough time to do the things you want to do, as well as the things you're obligated to do, doesn't happen by accident—it happens through planning and effective time management in all areas of your life. **If you choose to invest yourself in your research experience, you will, through necessity, develop and refine your time-management skills both in and out of the lab.** As a result, you'll spend less time cramming for exams and finishing assignments at the last minute and more time enjoying your overall college experience.

Learning to prioritize the activities that are important

With only twenty-four hours in a day, everything you do comes at the cost of doing something else. Every class you take, study group you join, YouTube video you watch, or volunteer opportunity you pursue requires you to take the time from somewhere. **When estimating the hours you have available for a research experience, you'll need to consider each activity and decide if what you get out of it is worth the time you spend on it.** On the surface, this may seem a simple task, but for many it can be incredibly difficult when faced with opinions from a well-meaning friend, family member, or classmate. *Learning to prioritize your time also means learning to trust your own judgment about how you should spend it.*

Intellectual Pursuits

Research has intellectual challenges beyond learning facts, understanding theories, and testing hypotheses. Pushing yourself to connect with your research on an intellectual level will lead to some of these personal benefits in the lab and outside of it.

Driving intellectual curiosity

In most research experiences, you can simply choose to go through the motions or make an intellectual investment in your project. In other words, you could choose to approach your research project as a series of tasks to be carried out and checklists to be completed. Alternatively, you could choose to understand why a particular technique or approach is relevant to your research, why the result of an experiment is important, or how your project supports the "big picture" of the lab's research program. **The path you choose is up to you, but doctors, researchers, and entrepreneurs who are interested the hows and whys are also the ones who discover or invent creative solutions to problems.** Use your research experience to nurture your intellectual curiosity (and creativity) early, so it is second nature by the time you need it in your life and your career.

Fostering creativity

Creativity is not limited to the arts. The experiential learning that accompanies an in-depth research experience will provide ample opportunities to develop creative-thinking skills — such as the ability to shift your perspective to consider new, possibly unconventional approaches to solve a problem or theorize an alternative yet equally plausible explanation for an unexpected research result. **Those who develop creative-thinking skills are less likely to stagnate in a professional position, more likely to find**

personal fulfillment and significance in their life, and more likely to introduce inventive solutions as needed in both. And, yes, unlocking your creative problem-solving skills will absolutely transfer to your life outside the lab and well beyond undergraduate work.

Sharpening critical thinking

Research teaches you how to evaluate the outcome of your experiments and formulate possible explanations. To be successful, it's essential to remove your personal biases from consideration, analyze the relevant facts, and set aside the inconsequential ones. This allows you to make conclusions and decisions based on the evidence and facts — not on what you want to be true. An in-depth research experience also trains you to notice when there isn't enough evidence to support a hypothesis or give weight to a conclusion. As you learn to do this through your research project, the ability will inevitably extend to your personal life. **You'll become a more discerning citizen and gradually begin to expect higher standards of evidence from your friends, coworkers, reporters, politicians, and Internet bloggers before accepting a statement as a fact.** Essentially, you'll be less likely to take someone's word for it and be more likely to recognize when a conclusion isn't supported by the evidence (or lack thereof).

Academic Advantages

Beyond professional and personal development, you gain academic advantages by participating in undergraduate research. These advantages will benefit you throughout your time in college, and some will continue to pay off after you graduate. Some advantages include these:

Engaging in the Scientific Process

As an undergraduate in a lab, you're immediately involved in the cutting edge of research. Being immersed in an undergraduate research experience will give you unmatched exposure to the process of science — how discoveries are made and confirmed and how information is obtained and synthesized. You'll be part of the joy of discovery and experience firsthand how new and old information is combined to construct new models of how processes work.

Making a Contribution or a Discovery

While it may be unlikely that you'll discover something absolutely new, **the discoveries you do make will likely help solve a problem or answer**

a question. Whether you answer a specific question about how two genes interact, or contribute insight into a developmental pathway, your contributions will add knowledge to the universe.

Experiencing Science in Action

A research project is the ultimate in experiential learning. Although you could learn about cell division from studying it in a lecture class, the key concepts will be unforgettable if your research project involves manipulating cell division and quantifying the phenotypic effects at the research bench. *Aristotle said it best: "For the things we have to learn before we can do them, we learn by doing them."*

Earning Course Credit

The advantages of taking research for credit vary significantly depending on your college and department. In some departments, undergraduate research factors into a GPA, and in others research for credit isn't offered. Some students register to get research on their transcript or to remain a full-time student while taking one fewer lecture class during the semester. Registering for research credit may also be required to meet certain academic requirements such as pursuing an honor's thesis or applying for certain fellowships.

Writing a Thesis or an Undergraduate Research Journal Publication

Writing an undergraduate thesis or publishing a paper in an undergraduate research journal is an achievement in itself. It will be included on your CV, and you'll have the personal satisfaction of funneling your hard work into a publishable manuscript. Beyond that, there may be other tangible rewards as well. In some academic programs, publishing an undergraduate thesis or research paper paves the way for graduating with honors, earning a degree with distinction, or eligibility for certain scholarships or awards.

Excelling in Coursework

Your research experience can give you an advantage in classes and supplement your classroom knowledge. **Depending on your research project, you might learn to use research tools, methods, or techniques that are covered in an advanced lecture or lab class.** To have even a basic introduction before studying them in a class is an advantage. For example, if you use polyacrylamide gel electrophoresis during your research experience, learning about it in your biochemistry class will be easy.

Preparing for the MCAT

Your research experience may include ample opportunities for you to critically evaluate the outcome of a technique or analyze the data from an experiment. **The more you hone your critical thinking and analytical skills, the better prepared you will be for tests that include these types of questions**. Many of my former undergraduates have commented on how useful the skills learned during their research experience were on the reasoning and problem-solving sections of the MCAT.

Exploring the Academic Bubble on Campus

If you're a member of a research lab, you'll be more likely to attend research seminars and symposia. Imagine no quizzes or exams—simply learning because you're genuinely interested in the topic. This will make an impact on you whether it's related to your current research or projected career path or is something that has personal relevance. The longer you participate in a research experience, the more inspiration you'll draw from attending such events.

Connections—Professional, Personal, and More

In a lab, every interaction is an opportunity to make a personal or professional connection. From technical assistance on a protocol, or instructions on how to use a piece of equipment, to learning about events on campus, getting career advice, or building lasting friendships with others, the connections you make with your labmates can be a source of both knowledge and inspiration.

Professional Connections with Other Undergraduates

Many of your undergraduate labmates will have career goals similar to yours. The older undergraduates will attend medical, graduate, or professional school (perhaps after a gap or personal year) or start in a professional position directly after graduation. These labmates will have advice and insight on application processes, classes, professors, volunteer experiences, club memberships, and other extracurricular activities.

When you solicit advice and opinions from a labmate, it's easier to evaluate the quality of the opinion or information, because you've observed the work ethic, academic commitment, and self-management of that person. If you know a certain shadowing or volunteer experience helped them determine that pediatric oncology is now their true path, or that a particular physical biochemistry class was too hard, you can use your personal observations to determine how much value to place in their advice and opinions.

In addition, labmates who will not be competing directly with you for a future professional position (because they are further ahead in the academic cycle or have chosen a different path) are more likely to share honest opinions about problems and pitfalls they faced or opportunities that they seized during their undergraduate years. They are also more likely to be straightforward if you ask them, "What are you absolutely glad you did?" And you should definitely ask!

Personal Connections with Other Undergraduates

On a personal level, you might form close friendships with other undergraduate labmates. Although you'll want to do the bulk of socializing outside of the lab, many students find that the personal connections they make with other lab members are substantial and can become lifelong.

Professional Connections with Gradate Students and Research Staff

Those who have "been there" can be a valuable source of advice as their perspectives looking back on their undergraduate experience are rooted in their successes and failures. Graduate students early in their graduate program aren't too far from their undergraduate days to remember the mistakes they made, the successes they had, and what they would definitely do again (or never again). **Professional researchers tend to focus on career building and know what opportunities helped them get ahead and which ones they wish they would have invested more time or effort in.** Members from either group are likely to share "insider" information that could help you excel at both the research bench and during your college career.

Connections with Your Research Mentor

The relationship you establish with your research mentor might be the most significant of all professional connections you make in the lab. In addition to helping you work through challenges at the research bench, they might advise you of opportunities outside of the lab that could be beneficial to both your professional and personal development. For some undergraduates, their research mentor becomes an important part of their overall success in college and in life after.

Connections with Your Professors

When I (DGO) was an undergraduate, a defining moment came during a conversation with a professor who asked me about my research. This was the first time I had a meaningful conversation with a professor about something other than lecture or course material. To get to discuss *my* research project was exciting, and although I knew it wasn't on the same level as a colleague or a graduate student, I still felt important and left the conversation inspired. As a professor now, the interactions I've had with my undergraduate researchers, and students I speak with at office hours, indicate that this is a universal, cross-generational feeling.

Participating in a research project—and understanding it enough to have a meaningful conversation with a professor about it—is a personal accomplishment. As a bonus, it also demonstrates that you're invested in your college experience. If the conversation is authentic (meaning you leave the BS in the hall), you'll likely set the stage for a letter of recommendation later.

Connections That Grow

When graduate students and professional researchers move on to their next career step, they could relocate across the country or even across the world. You might be able to visit former labmates for an additional research experience, or they might recommend you to one of their colleagues as their professional connections grow. If you take a personal year and decide to work as a research scientist, or decide that graduate school or a long-term professional research position is your career path, you will have more opportunities if you have already established professional connections.

Connections That Inspire

If you attend meetings and symposia, you will be exposed new ideas, concepts, and approaches in research. Some will be directly related to the research you do, and some will be tangential. **At any meeting, you're likely to learn more about how the research you do connects to an even bigger puzzle beyond your lab and to the world at large.** Of course you know you're answering an important question, but it's inspiring to talk with someone outside your lab who is excited to hear about what you've discovered.

Connections That Enlighten

Some scientific conferences include workshops for undergraduates on career development, academic opportunities, or research opportunities.

The workshops generally include information applicable to students at all stages and provide guidance for professional life after college. As a bonus, if you make connections with other conference attendees, you could learn about specific opportunities, internships, or a career path you've never considered (or even knew existed).

Connections That Take You Places

Once you've participated in a research experience, you'll be more likely to consider participating in additional research and research-related activities. You might investigate national or international research opportunities that include a summer, semester, or year abroad. Perhaps, as graduation approaches, you'll consider pursuing a master's degree during a personal development year in a place that you've never even visited! With research experience, and a specialized skill set, comes the knowledge and confidence that will make it more likely for you to consider (and pursue) new adventures.

Potential Financial Rewards

It's Nice Work If You Can Get It...

Undergraduate research can lead to financial rewards in several ways. You'll want to check with your office of undergraduate research for guidance on fellowships, scholarships, and internships, as well as search the Internet for research awards and programs sponsored by national or international organizations.

Awards and scholarships

These can range from money to purchase research supplies to travel awards to help defray the costs of attending a scientific meeting. Some awards include a stipend to help cover living expenses while visiting a lab that collaborates with your primary lab. In addition, some research symposia give awards for poster presentations or short talks that can be spent however the recipient sees fit.

Fellowships and internships

Paid research fellowships and internships are available at colleges, universities, national labs, government labs, centers, and institutes. **Many of these programs are open to undergraduates regardless of their home university.** National and international opportunities exist for enthusiastic, qualified undergraduates with solid recommendation letters. These can be ideal opportunities for an intensive research experience in a relatively short time and often include the additional benefit of a stipend and room and board.

Paychecks

In a research lab, as part of a semester experience, paid positions are more competitive than volunteer or research-for-course-credit positions. Without a fellowship or scholarship, a paid position is often more likely if a student has a financial aid award that can be used to help offset the salary paid by the lab. (Check with your college or university's financial aid department for eligibility.) Also note that some universities forbid earning class credit and a paycheck at the same time. However, it might be possible to be paid from a research fellowship or scholarship and earn class credit at the same time. If you do have the option of earning a paycheck or stipend, make sure it won't negativity affect your financial aid before you accept a position.

Teaching assistant programs

Once you have research experience, you'll be more competitive for summer opportunities to teach research at university-sponsored science programs for middle or high school students. These programs pay qualified undergraduates to serve as instructors for a few weeks during the summer. Programs range from teaching specialty topics, such as DNA-handling techniques or clinical-pathology techniques to full disciplines such as genetics or cell biology. Although the programs vary, responsibilities typically include hands-on teaching of lab techniques, a lecturing component, an advisory component, and a supervisory component. Most programs pay a stipend as well as room and board as financial compensation.

Non-research scholarships and awards

The same professional and personal strengths that you develop and refine in an undergraduate research position are valued by selection committees for non-research awards. Participating in undergraduate research also shows that you are investing in your college experience and are involved on your campus, which is helpful on scholarship, fellowship, and award applications.

Recommendation and Reference Letters

From Your Research Professor

One of the tangible benefits you can earn from a research experience is recommendation or reference letters. As a member of numerous faculty search committees, graduate admissions, and fellowship committees, I (DGO) have read countless letters. **By far, the ones that make the largest impact include direct observations by the letter writer about the professional skills and personal strengths demonstrated by the candidate.**

Participating in research can provide the opportunities to repeatedly demonstrate personal growth, professionalism, motivation, self-discipline, character, intellect, cultural competence, creative thinking—and so much more. **Essentially, if you can learn it, or refine it, you can demonstrate it.** Almost everything listed in this chapter can be included in a supportive letter on your behalf. However, as with the other benefits of research, the letters of reference or recommendation you earn will depend on your level of commitment, how much you invest yourself in your research experience, and the opportunities that accompany it.

As an aside, once you've established a solid rapport with your research professor, it's appropriate to ask for a recommendation letter when you apply for nonscience scholarships or awards.

From Your Lecture Professors

Everything you do to enhance your professional development through a research experience demonstrates your commitment to your education and future career path. When a professor agrees to write a letter of recommendation on your behalf, any professional development could be used to enhance her letter even if she is not a research professor.

One Note on Preparing for Letters

After agreeing to write a letter of recommendation for you, many professors will ask you to provide a draft letter detailing your successes and the qualities you demonstrated. For many undergraduates, this self-assessment is one of the most difficult "assignments" in college. Before you start writing each draft letter, review this chapter and highlight each personal or professional quality you demonstrated for each reference. (Use a different highlighter color for each reference.) This will make crafting each draft letter easier because you'll have a starting point of specific qualities to include.

CHAPTER 3

Understanding Research

What Is Scientific Research?

RESEARCH in the sciences can involve conducting or collating surveys, doing fieldwork, performing techniques in a lab, creating computer modeling, conducting clinical studies, contributing to theoretical work, and more. When it comes to laboratory research, which is the focus of this book, there are as many types of research experiences as there are labs and as there are cultures within those labs.

The Undergraduate Research Experience

There is no one-size-fits-all undergraduate research experience. Even students who work on similar projects in the same lab will have different research experiences. In essence, a research experience is custom to each undergraduate researcher based on their goals, dedication, mentor, and research opportunities within the lab. (This is covered in more detail in the lab culture section later in this chapter and in the section about setting realistic expectations in chapter 4 entitled Will I Like Research?)

Wet Lab vs. Dry Lab Research

A wet laboratory, or wet lab, is a lab equipped with liquid chemicals and reagents. The research carried out in a wet lab includes manipulation of samples through a variety of techniques that are done at a research bench. This is in contrast to a dry lab, where the work done is primarily data analysis, computer modeling, bioinformatics, and other scientific pursuits that do not involve liquids. Many research programs include both wet and dry applications. Although this book often uses research examples that might take place in a wet lab, most of the information and advice is relevant to all research experiences in STEM (science, technology, engineering, and mathematics).

Basic vs. Applied Research

Scientific research can be divided into two broad types — basic and applied. Basic research focuses on understanding fundamental processes; whereas, applied research is directed at solving a particular problem. **In practice, there is often significant overlap between basic and applied research.** For example, someone conducting basic research to understand the regulation of protein synthesis in bacteria may discover a new target for antibiotics. Therefore, the knowledge gained from basic research may have obvious and immediate applications. Similarly, someone carrying out applied research aimed at curing a specific disease might uncover a new signaling pathway, thus adding new knowledge about cell signaling.

Model Organisms and Why They Are Used

Although the biological world is composed of countless organisms, researchers conducting basic research typically work with only a few. These organisms are known as reference or model organisms.

Model organisms are chosen for study based on particular life history traits that make them amenable for lab research. For example, a geneticist might study fruit flies (*Drosophila melanogaster*) not because he is particularly fond of fruit or flies, but because fruit flies are easy to culture in the lab, have a short generation time, are easy to cross, and have a small number of chromosomes. However, if a scientist wants to study photosynthesis to understand the movement of electrons to create more efficient solar collectors, then fruit flies would not work because they are not photosynthetic organisms. A more appropriate model organism in this case would be the alga *Chlamydomonas reinhardtii.*

Research on model organisms is about understanding the unknown and the discovery of something new. Although this is not a complete list, other model systems scientists use in the lab include bacteria (*Escherichia coli*), yeast (*Saccharomyces cerevisiae*), plant (*Arabidopsis thaliana*), fish (*Danio rerio*), mouse (*Mus musculus*), and worm (*Caenorhabditis elegans*).

Don't Judge a Lab by Its Organism

Over the years, we've advised undergraduates who were hesitant to work in a mouse lab because they didn't want to handle animals, and students who didn't want to work in a plant lab because they didn't want a botany project. Although these are both understandable beliefs, they don't necessarily reflect the reality of what happens in a research lab. For example, **you could be in the lab four years and never see or handle the model organism the lab uses.** This is because much research is done on pieces and parts such as tissue samples, proteins, or purified DNA. **In reality, a project could involve cell culture, protein expression, cloning, assays, or microscopy, and you'd never know, based on the techniques, if the model organism was mouse, fly, fish, plant, worm, or algae**. In addition, you'll miss out on incredible opportunities if you have an incorrect assumption about what kind of research is done on a model organism. For example, a research project that involves protein folding could lead to downstream applications that are important in human diseases such as frontal temporal dementia or cystic fibrosis. A scientist could study these in a lab that works with mice, human cell culture, worms, flies, yeast, or plants. Yes, that's right, plants.

Therefore, you should not choose or rule out a research experience based on the lab's model organism, but instead base your search on whether the science inspires you. Don't close yourself off to opportunities without knowing what they hold. If you choose the right research position, it's the science that you'll connect with — not the organism.

How Labs Classify Themselves

Each professor uses a set of key words and categories to describe her research program. Sometimes the categories or disciplines overlap (this is called multidisciplinary). For example, a professor might use cell biology, genetics, microbiology, and bioinformatics to describe her research program or might simply use the term *cellular* or *molecular biology* to encompass all of the above. This will be important for you to remember when you consider which research programs to apply to. You won't want to immediately discount a research opportunity because it doesn't list a specific topic or discipline you're interested in because the professor doesn't list it as a specialty. Instead, you'll want to use her description to guide you to inspiration.

Research Is More Than Benchwork

When you think about research, what comes to mind? Scientists in white coats wearing goggles or gloves? People staring through a microscope or at a computer screen? For many, the initial response to "What is research?" is a variation on those themes.

Although it varies by discipline and specific project, there are seven broad, overlapping parts to a research project. Some parts, such as choosing how data are disseminated, are completed by the professor in whose lab the research is conducted. Other parts, such as updating lab notebooks, are basic expectations of all lab members.

Seven Parts to Research

Part 1: Background research

There are two types of background research: project background and technical background. Project background includes the basic information about a project, why it is important, and how it supports the overall research goals of the lab. Technical background includes the purpose of the techniques used in a project, knowledge of what happens during each step of a protocol, and how using a specific technique helps achieve project objectives. The more background a researcher understands, the more he can contribute to all parts of a research project.

Part 2: Planning an experiment

Experiments don't just happen. At the very least, they require a basic plan of what needs to be done and the best way to do it. A novice researcher won't be heavily involved in planning every experiment, but it's still important to understand the process. When planning an experiment, researchers keep the following in mind:

- What are the objectives and specific aims of the experiment?
- What techniques and specific steps will be used to complete the experiment?
- What controls will be included?
- What supplies, equipment, reagents, and samples will be needed?

Part 3: Preparing supplies, equipment, and reagents

After an experimental plan has been created, the next step is to gather the necessary supplies and reagents and prepare the equipment. This might include collecting samples, making solutions, preparing glassware, or programming a piece of equipment. Depending on the complexity of the experiment and the materials available, this can take a few minutes, a few hours, or a few months.

Part 4: Benchwork

This is what most people think of as research—wearing gloves, goggles, possibly staring through a microscope, and working at the research bench. The type of benchwork varies whether in a wet or dry lab, by academic discipline, and by the techniques used. Benchwork includes carrying out a technique or experiment or redoing one when a procedure fails.

Step 5: Updating a notebook

If it wasn't written down, it didn't happen. Faithfully recording procedures, observations, results, and data is paramount. Even the most successful experiment, with the most exciting data, is unusable in grants and papers if the details aren't correctly recorded in a notebook. Researchers spend a significant amount of time writing down what they plan to do, what they did, and what they learned.

Part 6: Analyzing results and data

For most researchers, this is the most rewarding part of research. This is when discoveries are made, hypotheses are supported or nullified, and new models are derived. Analysis of new results and data lead to exciting new questions to ask, directions to take, and experiments to plan.

Part 7: Dissemination of results

As important as data analysis is, it isn't worth as much if no one else knows about it. Communication of results includes everything from presenting at lab meetings, research symposia, and talks at conferences, to publishing in professional and undergraduate research journals and interacting on social media. Of course, not every research result will be disseminated through all these avenues, so each researcher discusses what is right for a project with their professor.

The Lab Environment

A research lab is a complex and often intimidating place—especially to a new member. Understanding the basic roles of the other lab members and the intricacies of lab culture will make it easier to navigate.

Lab Members and Their Roles

There are few areas where a "standard" for all labs exists. With the exception of the principal investigator, lab members and their roles fit in this category. The positions described in this section are generalizations of the roles and responsibilities held by some lab members in some labs.

The Principal Investigator (PI)

Note: The abbreviation PI will be used extensively throughout this book.

The PI is the overall lab head (aka "The Big Cheese" or "Fearless Leader"). Typically, PIs work sixty-five to eighty or more hours per week including most university holidays. The PI is an expert in her field of research, has a PhD (or other advanced degree such as an MD), and typically has several years of experience as a postdoc researcher. Most university PIs do not teach during the summer and instead focus on their research program — often without earning a salary. The PI's research responsibilities include determining the research priorities and overall goals of the lab, obtaining funding, publishing results, and making personnel decisions. The PI often teaches courses, writes recommendation letters for lab members (and students from classes), and participates in various committees in the home department and on the college and university levels. Additional responsibilities exist depending on academic discipline and college or university.

Professional research staff (PRS)

A member of the PRS could hold any number of positions or roles within a lab and might work full-time or part-time — it varies significantly depending on the position and the person's career goals. **If there is a long-established PRS person in the lab who holds the position of lab manager, he is probably second-in-command under the PI.** Individuals in this position typically include technicians, research assistants, research scientists, and others. A member may stay in a lab for only a few months or might be employed by the PI for five, ten, or more years.

Postdoc researcher. A postdoc's responsibilities can vary significantly among labs, or even within the same lab. A postdoc might work part-time to upward of seventy hours per week. If no one officially has been appointed to the position of lab manager, a postdoc might adopt the role. A postdoc's appointment might be temporary, with goals to publish papers, gain research experience, and develop a research program prior to applying for faculty positions, or it may be longer-term as described for PRS. Outside of the lab, a postdoc might teach lecture classes, write grant proposals, write manuscripts, create posters, and present research at scientific meetings. Postdocs hold a PhD degree, thus the abbreviation *postdoc* for postdoctoral.

Other paid researchers. A lab might have several employees who contribute to the success of the research program in a variety of positions and roles. The number of hours worked per week can range from as few as three to as many as forty. **Undergraduate students with paid research positions are often in this category,** as are employees who have some research experience but haven't been in the workforce long. Seasoned researchers who are not hired on a salary contract might also fit in this category.

Students

Undergraduate researcher. An undergraduate could be in the lab anywhere from three to more than fifteen hours per week depending on a variety of factors. **Responsibilities vary immensely depending on the lab culture, type of research, project, skill set, goals of the student, and commitment level.**

Graduate student. A graduate student's weekly hours vary considerably depending on a variety of factors. Research responsibilities include significant benchwork and planning experimental strategies with the goal of completing a thesis (unless pursuing a non-thesis degree), learning techniques, and preparing as much as possible for the next career step. A graduate student might be pursuing either a master's or doctoral degree. A graduate student might mentor undergraduate students and might have responsibilities that include aspects of lab management such as ordering supplies, making stocks, and helping to maintain equipment. Outside of the lab, a graduate student takes classes at the beginning of her graduate career and might serve as a teaching assistant throughout it.

High school students. Some programs place high school students in college and university labs to participate in a research experience. Typically, these programs are done over the summer and are an opportunity for PIs to give back to the community, as well as provide an inspiring research experience for the high school student. In some programs, the students are paid a stipend through a fellowship, and in others the student pays the university or college to participate. (PIs typically do not financially benefit from hosting high school students.)

Behind the Scenes

What the other lab members do all day

A research lab is a professional work environment. As an undergraduate, you're a part of it, but are not subject to the same stressors, pressure, responsibilities, or commitments that affect the other members of the lab. For example: Weekends? After 7:00 p.m.? University holidays? You might use these as opportunities to get caught up on sleep, take a road trip, or finish a class assignment. For a professional researcher or graduate student, these are opportunities to get more research done without the responsibilities of classes, supervisory duties, or campus parking issues.

A typical agenda for a professional researcher or graduate student might look like this:

- Research. Research. Research. A work schedule might be twelve or more hours per *day*, five or seven days a week
- The design, implementation, and data analysis of experiments
- Determining when to abandon an experiment or redesign it
- Planning undergraduate projects (or graduate student projects)
- Preparing stocks and reagents. The ones on the benches, in the cabinets, in the freezers, and in the refrigerators
- Reporting their results and data to the PI
- Writing papers and grant proposals
- Preparing to present at a scientific conference. Traveling to said conference. Stressing about the work to be done because they are at a scientific conference instead of in the lab
- Reading peer-reviewed journal articles
- Troubleshooting undergraduate projects, their own projects, other lab members' projects, and the projects of researchers from labs down the hall
- Microwaving food. Eating it over the sink. Making coffee. Drinking coffee. Thinking about making coffee. And drinking coffee... and microwaving food...
- Working on various aspects of professional development
- And much, much more

Your Research Mentor

In this book, mentor is defined as the person who oversees the majority of an undergraduate's research experience and training. Although it's impossible to characterize every mentor's role for every undergraduate researcher, typically a mentor will offer direction, support, encouragement, instruction, and guidance (although the levels vary significantly per the individuals involved). A mentor might also provide suggestions about opportunities that will enhance professional development and will care about an undergraduate's ultimate goal after graduation. A mentor might also suggest ways to use a college experience to accomplish personal goals and will encourage each student to make a full investment in their college experience.

The positive impact a mentor can make on an undergraduate researcher is immense — often the full effect isn't realized until a few years after graduation. This is especially true for those who start a research experience early and have the benefit of a mentoring relationship throughout their undergraduate career.

Who your mentor might be

Your mentor might be a graduate student, postdoc, another member of the professional research staff, another undergraduate, or the PI. Your mentor might be assigned by the PI, be the person who conducts your interview and invites you to join the lab, or be someone who volunteers to work with you after you've been in a lab awhile.

In some labs, a mentor isn't assigned at the start of a research experience, so you might learn how to follow a procedure from one person, make solutions from another, and keep a proper notebook from yet another. **In other labs, a mentor is determined before an interview is scheduled,** so you might have a go-to person from the start.

Why you might not work directly with the PI

Most likely, the PI will be considered your official mentor (especially if you take research for class credit), even if you primarily work with a different lab member.

The interaction a PI has with each member of her lab often depends on each researcher's position and training needs. Whereas initially an undergraduate will need help to understand basic research concepts, learn protocols, and properly operate equipment, a postdoc might need advice on advanced research strategies, reviewing a manuscript, or tips on applying to a faculty position. A graduate student might seek guidance from the PI on advanced research concepts, planning experiments, or mentoring an undergraduate.

For many PIs, project updates, results, securing funding, publishing, and mentoring lab members are fundamental objectives. However, when physically in the lab, a PI will often focus on moving her research program forward. This could mean synthesizing new information, discussing results and data, troubleshooting experiments, and discussing failed attempts and new directions. All these things involve training but on an advanced level. **Therefore, a PI isn't necessarily directly involved with hands-on training of new undergraduate researchers in research techniques.**

Most likely, as you demonstrate the value you bring to the research program, your interactions with the PI will increase in both number and substance. However, until then, understand that the PI will make an immense impact on your research experience and training because you are a member of her lab.

Lab Culture

A lab's culture is simply this: a mini-society consisting of a personnel hierarchy, a somewhat flexible system of rules and etiquette, and a basic understanding (although not necessarily agreement) among members on how the lab should function. Embedded in a lab's culture is everything and everyone connected to how the lab operates.

Specifically, some aspects of lab culture include the PI's management style, the system for training personnel (including undergraduates), how carefully safety regulations are followed, how basic lab chores are distributed, whether each member has a personal research space, how supplies and reagents are shared, how members of the lab interact with each other, and the opportunities for undergraduate researchers.

Within these categories are seemingly infinite variables, which is why no one-size-fits-all lab culture exists. Therefore, what is true in one lab is not necessarily true in another.

How lab culture is established

There are two significant influences on lab culture: the PI's management style and the other members of the research group. The two are intertwined, but the PI's chosen management style determines how much influence the other lab members have on the lab's culture.

What you need to know about the PI's management style before choosing a lab.

You don't need to know much other than the information presented next. Beyond that, you'll learn what you need to know when, *and if*, it becomes particularly relevant to your research experience.

In truth, it's nearly impossible to understand what the PI's management style is, and how you feel about it, until you're a member of the lab. Until then, you don't have a good way to evaluate others' opinions about it, or know if their experiences will mirror yours. As an undergraduate, the most relevant aspects of a PI's management style will be different from those most relevant to a graduate student or professional researcher. (Remember, you have different goals, responsibilities, and pressures.)

Also, regardless of the PI's management style, it's not relevant unless it has a negative impact on your happiness or prevents you from receiving the type of research experience you want. (As a graduate student, postdoc, or professional researcher, it absolutely matters from the start.)

It's true that both your happiness and success will be influenced by the PI's management style, but **the overall culture of the lab, your mentor's training style, and the training opportunities offered will have the greater, *direct* impact on your undergraduate research experience.**

Therefore, inquiring about a PI's management style won't be as enlightening as asking questions about the types of techniques you'll learn or the specific responsibilities you'll have as a lab member. As you spend time in the lab, the PI's management style will become important or irrelevant.

How lab members influence the culture

A lab culture is a complicated entity. Think of it as a living, evolving organism. Even the most involved PI does not spend 24/7 in the lab or make daily aspects of lab management his primary focus.

Therefore, how closely a research team follows the PI's guidelines, or chooses to get along with each other, will influence the overall lab culture. A solid research team will, for the most part, use the expectations established by the PI as a guideline to both their personal and professional conduct. In most labs, there is some flexibility in the areas that include day-to-day operations, and the chores (such as making media, defrosting freezers) are divided up somewhat equitably (except there is usually a lab member or two who choose not to participate on par with the others).

Lab culture and your research experience

Even though lab culture is connected to everyone and everything in the lab, some areas will have a more significant impact on your undergraduate research experience than others. For example, the application procedure won't matter much, but the time commitment and technical training opportunities associated with a position will matter significantly. Therefore, some aspects are only outlined as follows and are covered more thoroughly in chapter 4, Will I Like Research?

Keep in mind that none of the differences in lab culture are inherently good or bad. It's a matter of your opinion and your short- and long-term goals.

Fifteen ways cultures may differ among labs

1. **Experience requirement.** What counts as experience and how much, if any, is required varies. Not all labs require experience, and some mentors prefer undergraduates who don't have any.

2. **Application procedure.** Unfortunately, there isn't a standardized procedure to apply for a research position. Each lab member determines the selection criteria, application process, and interview procedure that works for them. Some use a quick selection process, and others follow a significant vetting strategy.

3. **Time commitment and schedule.** To join a specific research project, most mentors require a minimum time commitment in hours per week and semesters. How flexible that commitment can be, might or might not be negotiable.

4. **Types of positions.** How much time an undergraduate spends observing what others do, performing benchwork, or completing research-related tasks such as washing dishes is vastly different among labs. In some labs, an entire research experience might only include observing others. In other labs, an undergraduate experience could include a hands-on, independent project starting on the first day.

5. **Mentor's training style.** Each mentor determines the training style that works best for her. Sometimes that training style is compatible with an undergraduate's needs, and sometimes it isn't. Some mentors meet daily or weekly with an undergraduate researcher, and others are only available a few times during the semester. Some mentors are available on weekends or after 5:00 p.m., and will answer emails, texts, or phone calls quickly, and others basically disappear into the ether after the interview. Some mentors will walk undergraduates through new protocols step-by-step, and others will point to the pH meter with the instructions to figure it out. Even in the same lab, a mentor might distribute her time disproportionately among all the undergraduates she trains.

6. **Technical training opportunities.** The opportunities a lab provides in this category are extremely important to keep in mind in an

interview and when deciding whether or not to accept a specific research position. Certain projects require learning a set of standard techniques before conducting an experiment or collecting data. Other projects require the mastery of a single technique. Some projects will be a combination of wet and dry benchwork, and others will fit squarely into one of the two categories. A project might include techniques from a single discipline or encompass a broad range of cross-disciplines.

7. **Project contribution.** Some PIs and mentors expect undergraduates to participate fully in a project from start to finish, including the initial design strategy, benchwork, data analysis, and next research question to ask. Other PIs want undergraduates to assist experienced researchers or work on different aspects of several projects.

8. **Understanding the project background and applying classroom knowledge to research.** How much will be expected will depend on how important it is to the PI or mentor. Some believe that the type and amount of knowledge demonstrated by an undergraduate is directly linked to how motivated and creative a student is, and they mention so in recommendation letters. Others believe that it's less important for an undergraduate to take a holistic approach to research as long as a solid understanding of the project is demonstrated. Some PIs and mentors don't have expectations either way as long as the research gets done, and results or data are produced.

9. **Leadership opportunities.** Some PIs believe that it's important for undergraduates to help train others in basic techniques and take on a leadership role in the lab. Other PIs prefer that all training is carried out by graduate students or professional research staff to ensure that consistency is maintained.

10. **Responsibilities outside the lab.** In some research experiences, all work is completed in the lab during regular research hours. In others, work outside the lab can be substantial. For example, a PI may require you to write a research proposal, end-of-semester report, construct a poster, attend a symposium, read the PI's papers, design an experiment, or prepare for and attend group meetings, in addition to a regular research schedule.

11. **Course credit.** Whether or not a student is able, allowed, or required to register for undergraduate research varies depending on the PI and the department. Some PIs require registration during first

semester, and others only allow it after a student has demonstrated genuine interest in the research program.

12. **Salary.** Some PIs never pay undergraduate researchers to work in the lab, and others pay undergraduates to work on a specific project or as a general lab assistant.

13. **Undergraduates as authors.** In many labs, the PI determines authorships for publications. Some PIs do not put undergraduates on peer-reviewed publications; some do. Some put undergraduates on posters or abstracts, and other PIs decide on a case-by-case basis. (Keep in mind that for a variety of reasons not all research results and data end up in a publication.)

14. **Conference attendance.** The emphasis placed on attending conferences (an on-campus symposium or professional meeting) is directly related to how beneficial a mentor believes it will be to a student, and how a PI views undergraduate participation at conferences. Typically, a PI or mentor will encourage an undergraduate to attend such events if they believe the student will benefit, contribute, and be a consummate professional. Some PIs do not believe that large conferences benefit most undergraduate students and don't encourage attendance, although they don't discourage attendance either.

15. **Travel awards to conferences.** If an undergraduate has enough results or data to present at a professional conference, and is likely to use the conference to enhance their professional and personal development, most PIs will encourage the student to apply for travel awards either at the college level or from conference organizers. Some PIs will provide an additional travel stipend to help defray the cost of attending a conference.

CHAPTER 4

Will I Like Research?

Understanding Your Expectations — Your Strategy for Happiness and Success

WHETHER or not you'll like research has a lot to do with what you want to gain from a research experience, what you expect it to be like, and the research position you ultimately choose. Several years ago, I (DGO) interviewed an undergraduate who was unhappy with her research experience and in search of a new lab. She was smart, hard working, and genuinely interested in research, yet she was unhappy with her research experience because she had "only learned some techniques and wasn't really doing any research."

When she began to explain her lab responsibilities, I knew immediately that she was indeed conducting research and was building the research tools needed to conduct an experiment and test a hypothesis. For her, a simple but unfortunate misunderstanding was the stumbling block to happiness: her expectations were very different from the reality of the research experience she chose. To have been happy in her first lab, she either needed a project that included hypothesis testing from the start of her experience, or she needed to understand how her daily activities were directly connected to the "big picture" of the lab's research program.

By the time she interviewed in my lab, she was overwhelmed by her frustration and was almost ready to give up research completely. This one example highlights why realistic expectations from the start help prevent disappointment later.

What Influences Expectations and How Do I Know if I Have Any?

Expectations include the goals you wish to achieve through a research experience and what you imagine it will be like to conduct research. Expectations can be so subtle you're unaware of them, or so bold that it is nearly impossible to fulfill them. Either way, if you strive to understand what yours are, your research experience will be a more meaningful and rewarding use of your time. **Ultimately, you'll be happier and more successful if your expectations of what research should be closely match the experience you choose.**

What are some examples of expectations that undergraduates have?

Sometimes students, like the one mentioned earlier, start a research experience with expectations that lead to disappointment once they start. **The most common expectations and the reality in this category are these:**

Expectation: "I'm already an expert in the techniques from lab class, so research will be easy."

Reality: For most undergraduates, research turns out to be more difficult and more complicated than anticipated—even after completing lab classes. It takes time to adjust to a professional research lab and to gain the skills needed to be successful.

Expectation: "I only need research experience to get a recommendation letter, so the project doesn't matter."

Reality: Research is challenging. Being genuinely interested in the discipline, project, or the techniques is important; otherwise, meeting the challenges will be even harder. Plus, recommendation letters aren't part of a transaction for spending time in the lab, and some PIs only write letters for undergraduates who fully invest themselves in their research opportunity.

Expectation: "I want to be passionate about *something* so I picked research."

Reality: Passion can be ignited during a search, or it might be delayed until a student has acquired research skills and knowledge. For some, it might not happen if the undergraduate and project are not a good match. That is why the search and interview processes are so important.

Expectation: "I have a friend who loves research, so I want to do it too."

Reality: Each research experience is unique, because each undergraduate researcher is unique. Even when two undergraduates are in the same lab, and have similar projects, each will have a custom experience based on their mentor, goals, and the personal investment made.

Expectation: "Participating in research will make me feel important."

Reality: It takes time to feel confident in the lab and to connect with a research project. At the start of a research experience, most undergraduates feel awkward and experience more self-doubt than self-confidence. Undergraduates who push through the awkwardness are often rewarded with skills and self-confidence, but it takes time and patience with oneself during the transition.

However, having expectations isn't a problem — it's actually good. After all, it's those expectations that will guide you to the perfect research experience. The key is to recognize what you believe research will be like and reconcile it (as needed) with what the research experience you choose will *actually* be like.

You'll do both during your search for a research position and by asking pointed questions at your research interviews. To help you prepare, this chapter covers common concerns and expectations students have before starting an undergraduate research experience. **Essentially, it includes the information that we wish we had known before searching for our first undergraduate research positions.**

The Ideal Time to Start Undergraduate Research

No matter what type of research you want to do, or what your goals are, *start your research experience as soon as you can handle the time commitment without compromising your academics.* If you're able to start as a first-year student, all the better.

What Are the Advantages of Starting a Research Experience Early in My Undergraduate Career?

The more time and effort you put into a research experience, the more benefits you'll gain. (For details, review chapter 2, Why Choose Research?) In addition, some research fellowship or internship programs only accept students who are early in their college career. Others require research experience *and* that a student be early in their college career.

Starting sooner than later is also an insurance policy of sorts. If you decide that your first lab isn't the right one, you'll still have time to explore a different research opportunity. In addition, it takes time to earn a recommendation letter that will strongly support your application to graduate, medical, or professional school. Waiting until a semester or two before you need one is risky—especially if you aren't successful in the lab or don't have the opportunity to interact much with the PI before you need a letter.

Even if your ultimate goal is to write a senior thesis, you should still start your research experience as soon as you can. More time in a research experience will give you more opportunities to learn and recover from mistakes without the pressure of a thesis deadline looming. It will also give you more time to determine if you actually want to commit to writing a thesis, which can be as much work as a college course. If you decide to write a thesis, the more familiar you are with the research project and the background and significance of your project, the easier it will be to get the writing done.

Therefore, start your research experience as early as possible in your college career, to get the most out of it, and so you don't unintentionally disqualify yourself from opportunities.

Isn't It Better to Wait until I'm a Junior to Search for a Research Project, Because Upper-Class Students Are Preferred?

No. You should not put off the search for a research experience based on the belief that it's impossible to find a research project as a first- or second-year student. Even though some PIs or projects require upper-class standing, *many labs prefer undergraduates who are early in their academic career.* Still other PIs won't consider training a student who has junior standing. There are several reasons why someone might be unable to secure a research position, and academic standing is only one possibility to consider.

Upper-Class Students Have Completed Core Coursework and Lab Classes. Aren't These Important for Me to Understand the Research?

It's true that some research positions will require specific classes because they are needed to understand or make a contribution to a project. However, **many PIs have projects that require determination and hard work, not specialized knowledge or skills at the start.** So unless you read an advertisement for a research position that lists lecture or lab prerequisites, presume that they aren't required for a particular position.

What If I'm a Senior? Is It Too Late to Participate in a Research Experience?

No. It's not too late. Some mentors even prefer or require rising seniors for research projects. However, depending on the discipline, it might be more of a challenge to find a lab position and be able to finish a project so close to graduation. **The main challenge you'll face will be to find a mentor willing to choose you over someone who could be around longer.** From the start, you'll need to be exceptionally professional, direct, and prepared to demonstrate that you are genuinely interested in becoming a valuable member of the research team.

On a personal level, the disadvantage of delaying a research experience — beyond not getting the full scope of potential benefits or the same opportunities to enhance your resume — is the pressure you might put on yourself to make a research position work. If you do not like the research experience from the start, you'll probably need to decide if you can stick it out, or if you should abandon the idea of having a research experience completely. Your experience might be heavy on the observational side, with limited benchwork, or you might primarily assist others. Still, these are valuable opportunities to grow your resume, learn about science as a process, and earn a letter of evaluation.

Although it's better to start early, don't give up on a research experience even if you're a senior. Just prioritize your search as soon as you can.

What If My GPA Isn't as High as My Classmates'?

Do not let a lower GPA dissuade you from searching for a research position. True, there will be some positions that require a higher GPA than you have, so you won't be eligible for those. However, for many PIs, an undergraduate's GPA is irrelevant, and for others it's only one factor used to consider if a student is a good fit for a project. In addition, an in-depth research experience culminating with an epic letter of recommendation can significantly support a graduate or medical school application lacking in other areas. Even if your GPA isn't stellar, continue reading and commit to a search for your research experience. **There is no reason to disqualify yourself when plenty of PIs wouldn't.**

The Time Commitment

Ultimately, the research opportunity you choose must be compatible with your goals and the time you have available for a research experience. However, for most research projects, *a mentor will have a nonnegotiable minimum time commitment in mind* in weekly hours and semesters. Before you accept a position, it's essential to determine if the time requirement will work for you, or if you should pass on the opportunity.

How Many Hours Per Week Will I Be Required to Commit to a Research Experience?

The time commitment could range anywhere from three to fifteen or more hours per week during the semester to full-time during the summer months. **The number of hours varies based on the average time it takes an undergraduate researcher with similar skills and experience to complete a similar project.** (Some fellowship and internship programs require a near full-time commitment for several weeks during the semester or summer months. Also, some PIs require significant participation during the summer, regardless of fellowship status.)

Overall, the weekly time commitment during the semester will be influenced by the project objectives and the type of training needed to accomplish those objectives.

For example, if a project involves the expression, isolation, purification, and analysis of multiple proteins, the number of hours required will be more than a project that focuses on the sub-cloning of a single gene. Likewise, if a project involves learning multiple techniques, or even a single but particularly challenging one, then the time investment will be significant. However, if a project requires minimal training, or mastering a few easy-to-learn techniques, the required hours per week will likely be on the lower end.

In addition, for some research experiences, you'll be required to spend time outside the lab constructing posters, reading scientific papers, planning experiments, and attending meetings or research symposia.

How Will My Lab Schedule Be Determined and How Flexible Will It Be?

Your lab schedule is likely to change each semester along with your class schedule, extracurricular activities, and project's objectives. **The exact details of what your lab schedule will be are impossible to predict and will likely be covered at an interview.** However, some possibilities are that you might be required to adhere to a specific schedule, with a set number of hours per week and per lab session, or you might have significant flexibility in managing your lab time. Perhaps your schedule will be inconsistent: long days intermixed with short days but you won't know what kind of a day it will be until you get started. Maybe your schedule will have a significant unknown factor—you won't know if you'll have work to do until you arrive each day, or you might have the option of going in only when your experiment needs attending. **However, these factors will influence your lab schedule:**

Your mentor's work schedule and your training needs

If the bulk of your training will require your mentor to be present in the lab when you are, you'll likely need to fit your schedule around her availability. If the project requires minimal training or minor supervision, your lab schedule might be quite flexible.

Your track record in the lab

With time, and a proven track record of self-management and solid research skills, your mentor might agree to flexibility in your lab schedule as long as you meet project objectives, and no safety issues arise. This flexibility could include working at nights or on weekends, but the number of required hours will likely remain consistent. However, some projects leave little room for schedule flexibility, regardless of your abilities or track record simply because of the types of techniques or experiments done.

How Many Semesters Will I Be Expected to Participate in a Research Experience?

At the minimum, most mentors want a student to continue in the lab until specific project objectives or benchmarks are achieved. In essence, a mentor needs each undergraduate to produce results or data that offset the time and productivity lost to train the student. **For some mentors, that time is a single semester or a summer, and for others a year or more including a full-time summer commitment.** Typically, a mentor will state the time commitment she expects at the interview or beforehand in a position advertisement. If she doesn't, you should ask.

What If I Accomplish the Project Objectives and Want to Stay in the Lab Longer?

In research, as with most skill-centric activities, the more experience you have, the more success you tend to achieve. If your research project ends, and you're happy with the experience, definitely ask your mentor or PI if you can continue with a new project. The outcome is likely to be in your favor.

Because it can take both a long time and require significant effort to train an undergraduate researcher, the majority of PIs want to keep skilled, productive, and professional students around as long as possible. Therefore, even if your project ends, or your mentor has left the lab for a new adventure, most PIs will offer a new project if you've been a solid member of the research team.

However, some research experiences have a definite start and end date. For example, if a project involves collecting samples in the field during the summer, then the dates the work must be done are set. If you're interested in the next step, definitely ask if it's possible to participate in sample processing or data collection and analysis when the research moves into the lab.

What If I Don't Know How Long I Want to Participate in a Research Experience?

Right now, you don't need to know exactly how long you plan to participate in research, but it will make your search easier if you have a general idea. Fortunately, you've already laid the groundwork for determining this in chapter 2, Why Choose Research? When you start your search, you'll review the accomplishments you highlighted to help guide you.

What If I Want to Make a Shorter-Term Commitment Than the PI or Mentor Requires (Either in Weekly Hours or in Semesters)? Should I Accept the Position Anyway?

You should only apply to, or accept an offer for, a position if you believe you're likely to uphold the expected time commitment.

Some undergraduates believe that if they work hard, the PI will be impressed and overlook it if they bail before completing the agreed-upon time commitment. Other undergraduates believe that they will be able to negotiate a reduced lab schedule a few weeks in by claiming to need more study time.

It's true that most PIs would rather you quit than force yourself to stay in a research experience that wasn't the right one for you. It's also true that most will understand if you need to reduce your lab hours to keep your academics strong, because they want you to be successful out of the lab as well. However, that doesn't mean that reducing the time commitment you made will come without consequences, and in some cases you'll risk a very real penalty. (Especially if the PI suspects that you had no intention of upholding the commitment you made, the risk won't be worth it.) It's always better to be honest about the time commitment you can, and are willing, to make, and never agree to a commitment that you know you won't uphold.

Ten reasons that honesty is the best policy when it comes to your research time commitment

1. **You can find a project that fits the time commitment you're able to make.** Whether you want to participate in research for three hours per week for one semester, or a full-time summer before your senior year, you can find a project with the time commitment you want. Sometimes, a single lab will have several available projects with different time commitments, and you'll be able to choose which one you prefer.

2. **It's harder to care about your project if you've checked out before you've even started.** If you take a research position with the idea that you'll jump ship when something better comes along, or with the idea that you'll quit once you can list "experience" on your resume, it's unlikely that you'll be happy even in the short term. *If you're only at the search or interview stage and you're trying to think of ways to avoid the required time commitment, you haven't found the perfect research position* and need to keep searching.

3. **The hours might not be negotiable.** *In some labs, if an undergraduate isn't able to uphold the agreed-upon time commitment, they are dismissed instead of having their hours reduced.* Sometimes this is done to prevent a student from falling behind in classes, or so the mentor doesn't continue to train an undergraduate who is struggling to maintain their academic/life/research balance. Sometimes, however, dismissal is chosen because a certain number of research hours are needed to gain technical skills and make progress, and reducing that number will set the student up for failure. Therefore, when a mentor requires "X hours per week," he might be okay with regular time off to study or reducing your weekly hours — or he might not.

4. **The training plan is often designed with a specific time commitment in mind.** For example, if a project is to be completed over multiple semesters, there could be several distinct training aspects to it. The first semester might only include observing others, reading papers, or writing a research proposal. Alternatively, a project might be all benchwork the first semester, but only to learn skills, perfect techniques, or build the tools that will be used in experiments during the second semester.

5. **You might miss out on the interesting stuff.** Some PIs wait to put a new undergraduate on an independent project or something "interesting" until the student has demonstrated genuine enthusiasm, reliability, and a dedication to research. Although unpaid researchers don't have time cards, *if you miss enough hours, you might never get past the "test" phase to the interesting research phase,* even if you aren't dismissed from the lab. (The number of hours missed before it's an issue with the PI varies.)

6. **You'll risk being thought of as unreliable.** If you're in the lab less than the weekly hourly commitment agreed to at the interview, it's possible that labmates will need to cover your responsibilities, or parts of your project will be reassigned to another researcher. The thing is, it's unlikely that your mentor will discuss it with you—it will just happen. This can quickly become a vicious circle—you'll have less to do, so you reduce your lab hours even more, which leads to even fewer lab responsibilities. Once this pattern starts, it's easy to lose enthusiasm and hard to correct. Plus, it's *extremely difficult to regain your labmate's trust once you've been labeled unreliable,* and that can happen quickly—especially if their productivity is dependent on yours.

7. **You'll potentially sacrifice a recommendation letter.** Strong letters of recommendation are not guaranteed. If the PI suspects that you knowingly accepted the research position without the intention of upholding the time commitment, you might not end up with a letter at all. Also, *some PIs don't write letters for students who aren't around long enough (in hours or semesters) to demonstrate much professional or personal development.* So it's a risk to take a position knowing that you won't uphold the commitment—especially if your main goal is to secure a letter at the end. And there is no way to find out in the interview stage which type of PI is head of the lab and still be offered the research position.

8. **Lost opportunity of authorship.** If the project's objective is to produce specific results or data for a publication, and you depart the lab before the experiments are completed, or don't make enough progress, you could lose the chance of authorship on the resulting publication. The PI will likely assign the experiments (and authorship) to another researcher in the lab.

9. **You might lose a scholarship or fellowship.** Some awards aren't distributed until certain requirements are met, and other awards can be rescinded. Award requirements range from submitting a final report or paper, attending a scientific conference, or simply putting in the required number of hours at the lab by a certain deadline. For some awards, falling short of the requirements will automatically disqualify you from receiving the official award or the full benefits of the award.

10. **It's unfair to your mentor.** By planning an undergraduate research project, your mentor makes a commitment to you before you even start in the lab. As soon as you arrive in the lab, his efforts continue. Beyond teaching fundamental techniques, a significant amount of training goes into mentoring, advising, or working with an undergraduate researcher. Teaching someone how to think like a scientist and guiding them through techniques and experiments takes time. The development you receive from your research experience comes from the effort you make, but it also relies on a mentor who is equally committed to you and your success. Just as you hope that your investment in research will pay off with professional and personal development and a letter of recommendation, your mentor hopes training you will pay off in results or data that increase his productivity. It's rude to accept a position if you have no intention to uphold the commitment necessary to make the training you receive help your mentor get ahead. Especially if your mentor chose you over a student who would have made the commitment in good faith.

Does That Mean I'm Stuck If I Accept a Two-Semester Project, and I Decide I Don't Like It, or I Become Overcommitted with My Academics and Want Out or Need to Reduce My Hours?

No. No one will expect you to continue with a research experience if you don't like it or if it might compromise your academics. When an interviewer says something similar to, "I'd like you to make a one-year commitment," she means, "I'd like you to make a one-year commitment *in good faith.*" This means, she expects you to accept the research position only if you are genuinely interested in it, and if you believe that you'll uphold the time requirements barring unforeseen circumstances. If you find that it's not the right experience, you can make a professional departure. You just need to be aware of the information just presented so you understand the importance of making a commitment to a research position in good faith.

The Experience Paradox

Nothing is more frustrating than to be told that you need experience to get a research position, when you can't get experience without having held one! Fortunately, *experience* is a broad term; there are research positions that do not require it; and some mentors even prefer undergraduates without it.

How Much Research Experience Will I Need to Get a Research Position?

Some mentors require a specific skill set or research experience to be considered for a position. Experience in those cases could be a specific lab class, working knowledge of a particular technique, or experience in a professional research lab. Even in the same lab, there could be variation. One mentor might require experience with microscopy; whereas, **another might prefer to teach her undergraduate everything they need to know.**

So where does that leave you? Essentially, it's all up to the mentor. So, for some positions, yes, you'll need experience to be considered, but for other positions, an inexperienced student who demonstrates a genuine interest in the research project and the right schedule will fulfill all the requirements. **In many disciplines, the majority of undergraduates start their research experience with only lab-class experience.**

What Is It Like to Start in a Lab without Research Experience?

Surprisingly, it's similar to starting in a lab *with* previous research experience. Everyone feels awkward in the beginning. It's a new environment, new procedures, new expectations, and new people. However, if you have little to no experience, you won't be expected to be an expert or know how to keep a proper lab notebook. You'll probably find labmates willing to give you extra guidance in basic protocols as long as you work hard, follow instructions, and focus on learning. **As with any new situation, you'll adjust and it will get easier with each day.** You'll also generally find more than one person who will make you feel welcome, as you start to feel like a member of the lab.

How Difficult Will It Be to Learn the Research Techniques?

Unfortunately, there is no easy test to determine how hard it will be to learn the techniques and gain the skill set needed for a given project. Research, depending on the discipline, can have a steep learning curve — more for some than others. What we can tell you is that we've known very few undergraduates who were unable to learn the fundamental techniques as long as they were willing to put in the effort and follow instructions. **Overall, it's likely that your particular progress will depend on the following**:

- Your dedication, self-motivation, and perseverance in acquiring the fundamental bench skills
- Your ability to comprehend and follow instructions, and the self-reliance to carry them out with minimal supervision
- Your patience, self-discipline, and perseverance in conducting repetitive work
- The time commitment you make to your research experience

What If I Make Mistakes?

One of the few guarantees about research is this: you will make mistakes. After all, you'll be learning new procedures, acquiring new skills, and discovering the boundaries of your knowledge. Making mistakes is simply an integral, unavoidable part of this learning process. **However, the biggest mistake you could make is to let fear stop you from pursuing a research experience.**

As you gain experience, your mistakes will decrease, and the success you celebrate will increase. On your first day, you'll know the least amount of information and have the least number of skills. Each day after, you'll know more and be a little more confident than the day before. Until one day you'll have an incredible realization along the lines of, "I got this," and it will be the best research day ever.

Lab Classes

No matter what your research project is, there are advantages to completing lab classes prior to starting a research experience. *The more exposure you have to research, in any form, the better off you will be.* Nonetheless, if you don't have lab classes as experience, don't let it discourage you—many mentors don't require specific lab classes before starting a research experience.

I've Taken Lab Classes, Am I Prepared for Research?

In every lab class, you're introduced to a type of research. Of course there is considerable variability among lab classes—different disciplines, different techniques, different emphases—but all lab classes are valuable even if the advantages aren't immediately clear.

There are advantages to completing lab classes before starting a research experience. At the very basic level, you'll acquire a familiarity with some lab equipment and probably learn the differences among items of glassware such as a beaker, flask, and graduated cylinder. You might learn what the meniscus is and why it's relevant when measuring liquid, perhaps how to adjust the pH of a solution, or the proper way to dilute concentrated stock solutions. Some lab classes emphasize the scientific method, the importance of a good notebook, and experimental design. These all serve as a foundation for you to build on and help you transition into a research lab faster.

A lab class also can help you decide if you like (or dislike) a set of techniques or a type of science. This knowledge will be invaluable when you start your search for a research lab. Over the years, several undergraduates have joined our lab after finding inspiration from the techniques learned in a general biology, genetics, or microbiology lab course. Conversely, we remember one student who remarked that the techniques he learned in a biology lab were "a tedious waste of time," so we knew that he wouldn't be a good match for our lab.

Will Lab Classes Teach Me the Techniques I Need to Do Research?

Maybe. It will depend on what techniques you learn in the lab classes and the research project you join.

If you learned the basics of PCR in a lab class, and your research project includes this technique, or you handled bacterial stocks in a microbiology lab and your research project involves bacterial manipulation, your experience could be directly applicable. If you took a lab class that focused on sequencing the genome of an organism, then you would be prepared for a variety of research projects that incorporated similar techniques. However, if you took a chemistry lab and your research project is about wildlife ecology and conservation, although still valuable, you'll probably find that the techniques you learned in the chemistry lab aren't directly applicable to the project.

If I Learned a Technique in a Lab Class, Will I Be Able to Do It by Myself in a Research Lab?

It depends on how much of the technique you completed in the lab class and how much the teaching assistant did for you. In a lab class, several steps of a technique or experiment might be done by the teaching assistant before you arrive or after a lab session is over.

Therefore, you probably won't be an expert after learning a technique in a lab class, but you might have a significant foundation that will help you succeed with benchwork related to your project. This is important to keep in mind, so you remember to be patient with yourself if your first few weeks in a research lab are a little bumpy. You don't want to be disappointed if it turns out that you aren't as prepared as you had hoped you would be. It's similar to studying a foreign language before traveling abroad. You quickly realize that conjugating verbs and stringing words together to have a meaningful conversation is more difficult than anticipated, but much easier than if you didn't have previous language classes.

Knowing the name of a technique and being familiar with it isn't the same as knowing how to execute it from start to finish with a protocol that is slightly different from one you used in a lab class.

In the Lab Class I Took, the Experiments Failed Sometimes, So I Didn't Learn How to Do a Technique or an Experiment That Was Listed on the Syllabus. Will That Be a Problem If the PI Only Accepts Students Who Have Completed the Lab Course?

All PIs know that sometimes research fails—even in a lab class. Most PIs don't expect an undergraduate to be an expert after taking a lab class, as much as to have the basic knowledge or technical proficiency the class provides. If more than that is required for a research position, you'll find out at the interview.

However, there is an unexpected bonus if some techniques or experiments failed in your lab class. As odd as it sounds, to have firsthand knowledge that even the most well-planned experiment or technique can go awry is an advantage. **Undergraduates who start a research experience with this understanding are more resilient when faced with failures in**

their research, and that makes the transition to a professional research lab easier.

How Do Lab Classes and Professional Research Labs Differ?

It's impossible to characterize all the potential differences among all lab classes and all research labs. Lab classes vary from those that focus on general or introductory techniques, those that integrate several disciplines (such as biology, chemistry, and physics), or those that focus on a single research objective to allow the students to write a manuscript and submit it for publication. Plus, as you know, each research lab is a unique entity with a unique culture. However, there are some general differences among lab classes and research labs that often catch new undergraduates off guard.

Ten potential differences between the undergraduate experience in a lab class and the undergraduate experience in a professional research lab

1. **Scheduling lab time.** Out of necessity, most lab classes have a start and stop time for each session. If your lab class is scheduled for noon to 3:00 p.m., you're out the door by 3:00 p.m. so the next class can begin their session. In a research lab, your schedule is dictated by the procedure you do each session. For example, if you're scheduled from noon to 3:00 p.m., and your experiment isn't finished by 3:00 p.m., for whatever reason, you stay until it's at the correct stopping point, regardless of what the clock says. Quitting an experiment early can be expensive in time and lab resources, especially if a labmate's experiment is based on yours.

2. **Moving forward.** In a lab class, you might wrap up a set of techniques or experiments in a session or two, and you might never do the same technique twice. In a research lab, you might use the same set of techniques for several days or weeks (or longer) before moving on to something new, or you might learn a core set of techniques that you use throughout your experience.

3. **Preparing supplies and equipment.** Locating what you need is easy in a lab class where reagents and supplies are often at your research bench or close by when you arrive. Incubators are set to correct temperatures, and stocks and reagents are ready to use. In a research lab, you'll need to learn where the community supplies and reagents are kept, retrieve them when needed, and put them

back when done. You'll likely learn to prepare many of the reagents you need starting with the solid chemical, a scale, and the correct type of water. In addition, you'll likely be responsible for programming equipment and possibly maintaining personal stocks of the organism you use such as cell lines or seed, worm, bacterial, or fly stocks.

4. **Designing experiments.** In a lab class, teaching assistants and professors work hard behind the scenes to make techniques and experiments as fail proof as possible. Experiments are designed with careful thought given to the amount of time available in a lab session and the presumed technical skill level of the students. In a research lab, although a student's level of expertise is sometimes taken into account, experiments are designed to meet project objectives, and techniques can be especially challenging until certain skills are acquired.

5. **Keeping notebooks.** In a lab class, your notebook is turned in for grading, but often it's yours to keep at the end of the semester. In addition, you'll receive instructions from the lab manual or the teaching assistant as to what information needs to be recorded. *In a research lab, "your" notebook doesn't actually belong to you—it belongs to the college, university, or your PI,* and some PIs have a strict policy that it never leaves the lab. Your research notebook is as essential to your mentor and the PI as the results and data you produce. In addition, the information you need to record in your research notebook isn't always immediately obvious. Often, guidelines, rather than specific instructions, are offered, and confusion as to what needs to be recorded accompanies the start of each new research project.

6. **Working outside the lab.** In a lab class, you'll likely be required to read a protocol or an experimental overview before class, write lab reports, and possibly prepare a poster or do an additional out-of-class assignment. In a research lab, you might be expected to read some of the PI's papers, learn background material about techniques or your experiments, create and present a poster at a meeting or symposium, learn about your labmates' projects, or design an experiment.

7. **Knowing the plan.** In a lab class, the syllabus and lab manual make it easy to learn the objectives of each lab session before arriving. In

a research lab, your activities for a particular day might be unpredictable. When you arrive for the day, your cultures might not be ready for processing; your cell lines might be contaminated; or your mentor might have you work on an unexpected task. Your research strategy might even change during the lab session. With a research project, being prepared also means being prepared to change your plans when needed.

8. **Getting feedback.** In a lab class, it's pretty clear-cut: the syllabus covers your responsibilities, and there are graded assignments such as a practicum, a notebook, and quizzes. You can refer to the syllabus to calculate your grade during the semester. In a research lab, your mentor's expectations often won't come with such direct (or frequent) feedback—and *even a letter grade of A at the end of the semester doesn't guarantee an outstanding recommendation letter.* It also doesn't automatically mean that you've met your PI's or mentor's expectations—only that you've met those required for a letter grade of *A*.

9. **Obtaining help.** In a lab class, you can rely on a professor or teaching assistant (or two) to quickly help you when you get stuck or need clarification. In a research lab, your mentor might not be available to help you throughout your lab session, or her time in the lab might not even overlap with yours. You might need to postpone an experiment or set up an appointment with your mentor outside of lab hours if you get stuck.

10. **Repeating experiments.** If an experiment or technique doesn't work in a lab class, the teaching assistant might let you borrow supplies or notes from another student or group, or you might abandon the experiment and move to the next one. In a research lab, you'll likely need to start the procedure over if you are unsuccessful the first time. Or the second time. Or third. It might take multiple attempts over several weeks to achieve the desired result before you can move on to the next step.

Lab Positions for Undergraduate Researchers

What Will I Do All Day?

This is the question asked most frequently at interviews—and for good reason. How you spend your time in the lab is the core of your research experience. How much time you spend planning versus conducting experiments, and setting up versus cleaning up, will be determined by your skill level, training needs, project objectives, and the responsibilities of the lab position you accept.

Essentially, there are three ways for an undergraduate to join a lab: as an observer, a researcher, or a general lab assistant. These aren't necessarily distinct positions, as there can be significant overlap among them.

Observer

If you accept this position, you'll observe what others do, ask questions when it's convenient for them, and possibly have opportunities to do some hands-on research tasks. **Your goals will be to learn everything you can, determine if you'd like to stay in the lab long term, and demonstrate an interest in the research program.** At the same time, the other lab members will determine if you're a good fit for the research team and if your enthusiasm is genuine.

Researcher

As a researcher, most days will probably involve a combination of doing benchwork, updating your notebook, and contributing to the overall operation of the lab by completing research-related tasks. (Review the Seven Parts to Research in chapter 3, Understanding Research, for additional possibilities.) You may also spend some time observing other researchers. You might work on an independent project or work as an assistant to a seasoned researcher. With a research position comes the fun, stress, and responsibility of adjusting to the lab culture and learning new techniques at the same time.

General lab assistant

This position is sometimes referred to as a dishwasher or a lab assistant position. Usually, this is a paid position where primary responsibilities are centered on completing tasks that help keep the lab running such as washing lab ware or making media. **If you want to use this as a steppingstone to a research position, confirm with the PI that it will be possible before you accept the position.** Some general lab assistant positions don't include the opportunity to advance to a research project.

Why Would I Join a Lab as an Observer Instead of a Researcher?

Do not underestimate the significance of an "observe to learn" opportunity. **This is a common strategy PIs use to screen potential researchers, while**

introducing them to the inner workings of the lab. Understand that a PI will only offer this opportunity if she believes you might be an asset to her research team. It's up to you to demonstrate that her presumption is correct and to ultimately secure a research position.

For you, this opportunity can only pay off. Because once you're in the lab, even for a few minutes, you've learned something. You've learned about the lab's research focus, how experiments are carried out, details about specific techniques, or the lab culture. If you have the opportunity to help with an experiment, you'll gain hands-on experience related to what you'll do if you stay in the lab as a researcher. **Most importantly, you'll have the opportunity to evaluate if you're in *the perfect lab for you* before you make a formal commitment**. Essentially, you'll gain almost all the benefits of being in the lab, without the full responsibilities or pressure of being a researcher or general lab assistant.

Even if you decide not to continue in the lab after the probationary period, the opportunity to observe, learn, or gain hands-on research experience is an unparalleled advantage. Experience, knowledge, and training are benefits that become advantages — even if they are from a short-term experience.

What Is It Like to Join a Lab as a General Lab Assistant?

At the start, you'll most likely wash lab dishes. A lot of dishes. More dishes than you can stand. If you do that well, you might learn a few research-related tasks such as how to rack pipette tips, pour gels, make media, or prepare stock solutions. Then you'll probably wash more lab dishes. After you demonstrate reliability and consistency, you might learn a research technique either by observing someone or assisting them with benchwork.

How Will I Get "Promoted" to a Researcher Position?

If the PI agrees that the general lab assistant position could be used as a stepping-stone to a research position, then what you'll need to do to make that happen is simple. **Do everything you're instructed to do and do it well. Do it well every time. Every time. Every. Time.** No matter how boring the task. If moving into a research position is possible, then your performance as a general lab assistant will be a test. It will be a test of how well you follow instructions, follow through with tasks, your ability to work with others, your work ethic, and your self-reliance. Every research project has at least one boring phase and several boring steps. If you demonstrate to the PI that you will do a quality job on the most boring task, even after you've done it a thousand times, she'll know that you won't slack when you hit a boring, repetitive part of a research project. Essentially, she'll know that you're a good candidate for a research position.

However, if you complain about the work or show up late, irregularly, or without enthusiasm, or you do a shoddy job, then you'll prove to the PI that she made the right decision not to put you on a research project. In addition, it won't be long before you're dismissed from the lab, because **PIs have a tendency to become annoyed when the simple things aren't done well, or when they pay for subpar work.** We assure you, a PI will not move you onto a research project if you don't meet her expectations as a dishwasher, or if you don't feel that your talents are being utilized in a general lab assistant position.

Over the years, we've hired several undergraduates as general lab assistants. In many cases, a student excelled, was "promoted," and stayed in the lab long term as a researcher. If you deliver a solid, consistent performance as a general lab assistant with the understanding that you could be promoted to a research position, it will probably happen within the first semester, or before the start of the second.

Which Position Should I Pursue if I Want a Paid Lab Position?

With research budgets shrinking and the cost of research increasing, paid positions for undergraduate researchers are less plentiful than in the past. Some PIs will hire an undergraduate as a researcher, and others will only hire a student for a general lab assistant position. It's safe to bet that no PI will hire someone as an observer.

Unless you've been awarded a research fellowship, a general lab assistant position is typically your best bet for earning a paycheck. If you have a financial aid package that can be used to supplement a portion of your wages, this might make it easier to find a paid position. Check with your financial aid office for your options.

Why Would a PI Hire a General Lab Assistant?

To the PI, the advantage of hiring a general lab assistant is this: **Every lab has a set of chores no one likes to do, but are essential to keep the lab running smoothly.** Dishes need to be washed, media needs to be made, and glassware needs to be sterilized. Even in labs with access to an automatic dishwasher, someone has to load the glassware, run the cycle, and put the clean dishes back in the cabinets. No matter how efficient all the lab members are with these tasks, there is never enough time to get them all done. Hiring a general lab assistant to do those chores benefits the entire lab as soon as the person starts.

Which Type of Position Is the Best?

The best position is the one that lets you get your foot in the door and gives you the opportunity to learn. Some labs will only bring in new undergraduates as observers or lab assistants, and some labs always start students off as researchers.

What Happens If I Don't Like the Lab after I Join?

You'll use the search, application, and interview strategies in this book to increase the likelihood that you find the perfect lab the first time around. However, **if you join a lab that is incompatible with your goals or happiness, don't panic—you don't have to stay in a research experience that is wrong for you.** Even if you registered for class credit, and it's past the drop-and-add deadline, keep in mind it's not for the rest of your life—it's just until the end of the semester. If you didn't register for class credit, and you don't like the research experience, you can make a professional exit sooner.

And remember that there is value in trying something new, even if the main thing you learn is that you don't like it. *Being able to cross items off your things-to-try list is as important as putting items on it.*

Choosing a Research Project

Often, a lab will have multiple ongoing projects, each of which may have one or more subprojects that support the lab's research focus. For example, a lab's overall objective might be to understand how microtubule organization controls plant cell wall expansion. One project might use *in vivo* localization of the microtubule severing protein, katanin, during epidermal cell expansion. A parallel project might be the analysis of mutations that enhance a specific, well-studied katanin mutation. A subproject of that might be the map-based cloning of an enhancer mutation, and another might be the transformation of an enhancer mutant with a GFP-tagged katanin fusion.

Each of these projects and subprojects has the ultimate goal of providing new knowledge about how microtubule organization controls plant cell wall expansion. **As an undergraduate, you might participate in multiple projects, observe or assist on a single project, or be assigned your own subproject as independent research.**

What If I Want to Ask My Own Question and Design My Own Project?

Answering your own question, or investigating a problem that intrigues you, can be an incredibly rewarding experience. It's exciting to be an "undergraduate PI" with the challenges and responsibilities that accompany this role. **Although you might find a PI willing to sponsor your research from the start, it's likely that you'll be required to volunteer in his lab for a period of time first.** After which, you'll probably discuss your research ideas with him, do background research (in the library or online), and write a proposal that outlines your project objectives and the techniques you'll use to accomplish them. In some cases, the PI will be listed as your official mentor, but your project might still need approval by an administrator such as a director of undergraduate research in your department. Although you'll work under the guidance of a PI, choosing this route will take a special amount of dedication. Therefore, it's important to thoroughly consider if you're prepared for the level of independence and self-reliance that this approach requires, before you start down this path.

Do All Labs Have an "Ask My Own Question" Option?

No. In some labs, there is no chance of designing your own project—even if you start in the lab as a first-year student and stay until you graduate. **In some labs, even graduate students, postdocs, and professional researchers don't choose their own research questions,** even if they design and conduct the experiments to answer them. However, many labs have a happy medium that allows a significant amount of self-reliance and self-directed work.

What Is a Happy-Medium Project?

Many PIs use the terminology "independent project" for research experiences that fit in this category. **It's a balance between having complete control over your own project and being an assistant on someone else's project**. It's likely that once you acquire certain skills, demonstrate self-reliance, and achieve your initial benchmarks, you'll be given autonomy with respect to which research strategies to use, or which lab subproject to work on. You might also have input as to the direction your project will take. In a sense, this is answering your own question—it's just within the framework of the lab's research focus.

If I Don't Want to Design My Own Project, Will It Be More Difficult to Find a Research Position?

No, it's probably easier. Working on a project designed by the PI or on a project that falls within the happy-medium category is common. Matching a student to a specific project depends on the lab's current research goals, the student's technical and analytical skill set, and the amount of time the student is able to dedicate to research. (Remember, some projects have a nonnegotiable time commitment.) Nonetheless, most projects involve hands-on research experience, result analysis, or data collection. Your involvement level will depend on your project objectives, responsibilities, and how the project fits within the overall goals of the lab.

How Are Projects Prioritized?

Ultimately, the PI sets the lab's priorities, and those priorities can change depending on a variety of factors. Even within the course of a few months, one project may become irrelevant, while another becomes the new top priority. New data, a disproven hypothesis, or impending grant or paper submissions could influence why some projects are set aside while others are given higher priority. Even project objectives can change. Because research is about discovery, new results and data often produce a new set of questions to be answered. With that come new directions, possibilities, and decisions to be made. On occasion, a project may even be abandoned.

What If I Want to Work on a Different Project after Being in the Lab for a While?

How easy it will be to change projects, or if it will be possible at all, will depend on the reasons you want to switch and other circumstances. Although most labs have multiple active projects at any time, that doesn't mean the PI will automatically support a change, or that someone will be available to train you on something new. A PI will carefully consider the reasons you wish to make a change and probably consult with your mentor before making a decision.

Will I Work as Part of a Team or on My Own?

You will always work as part of a team regardless of the lab you join and whether or not you design your own project or work on one created by the PI. However, your project might or might not be an independent study. This is both lab-dependent and project-dependent. In some labs, a single project is divided among several researchers, and each person works on a different part. Within that framework, each part could be independent of the next or could be intertwined so part A is completed by one researcher then part B is completed by another. Other possibilities include a project with significant collaboration with a specific member of the research team (such as your mentor), or a project that requires you to work mostly independently with occasional consultations from a seasoned member of the lab or the PI.

To Register or Not to Register

Whether or not you register for credit for research is only partially up to you. Not all departments offer course credit for undergraduate research, and some PIs don't permit students to register for credit until they have demonstrated initiative, dedication to the research experience, or a particular skill set. However, some PIs require undergraduates to register for credit starting with their first semester in the lab.

Should I Take Research for Course Credit?

Before you register for course credit, it's important to have a solid understanding of what will be required and how registering will affect you academically. Therefore, you'll want to consider several factors and consult with your academic advisor before making the decision.

Start with the basics

You'll need to know if registering for undergraduate research will count toward your electives, and how many credits (if any) will count toward your major and degree. You'll also need to know if your research grade will factor into your GPA. Sometimes, undergraduate research is graded on the *A, B, C* scale or taken as pass/fail. Alternatively, research credits might be listed on a transcript but not graded.

Learn the department's requirements

Most departments that offer course credit for undergraduate research have a learning contract that covers the official requirements. **Typically, there will be a set number of hours a student is required to spend in the lab per registered credit hour. There also may be additional requirements.** For instance, some departments require students to write a research proposal about their project or a paper that summarizes their research results. If the department hosts a year-end research symposium, creating and presenting a poster might also be required.

Understand the PI's requirements

Sometimes PIs have additional requirements not covered in the departmental research contract. Although it varies, it wouldn't be unreasonable to be expected to turn in your notebook, or make a backup copy, clean your research bench space, and make sure that your research samples are properly labeled and stored prior to a grade being assigned. In actuality, these are likely to be expected regardless of your registration status. Some PIs also require a specific number of hours per week regardless of how many credit hours are assigned for research course credit.

After you know all the requirements, evaluate your previous semesters. Consider if any of the following are true:

- Did I drop or withdraw from classes after the initial drop-and-add deadline because I was overcommitted or ran out of time to study or complete class assignments?
- Was I stressed the *majority of the time* about not having enough time to complete assignments, study, or spend time with my friends?
- Did I spend *the majority of my study time* cramming before an exam or rushing to finish most out-of-class assignments before the deadline? Did I ask for extensions on assignments or papers or pretend to be ill to postpone an exam?
- Do I regularly feel overwhelmed with the amount of information from my classes, but I can't remember much of it or aren't sure that I'm learning anything?

If you answered yes to any of these questions, avoid registering for course credit the first semester you participate in research. As you gain research experience, by necessity, you'll develop solid time-management skills and learn to prioritize your time. Once you have these down, participating in research for course credit will be much less stressful.

You should also make an appointment with the campus counseling or study center to learn additional tips and tricks that will help you manage stress and create a solid academic/life balance.

Why Research Positions Are Competitive (and What You Can Do about It)

Depending on your major and career path, you may have already heard numerous times how important it is to get involved in undergraduate research. But if an undergraduate research experience is so important (and it is), and has so many potential benefits (which it does), why it is so difficult to find a project compatible with your long- and short-term goals? In other words, why are research positions competitive?

Four reasons research positions are competitive

1. **The number of students who want to participate in undergraduate research typically outnumbers the available positions.** In essence, there are a limited number of faculty members in each department (much smaller than number of students majoring in the subject or searching for positions), and of those faculty members, only some mentor undergraduate researchers. This will become frustrating if you spend too much time thinking about it, so instead focus your attention on other matters that you can control.

2. **Some research positions require previous research experience, an established skill set, prerequisite courses, or a specific academic standing.** If you don't have the required qualifications, it's unlikely you'll be considered for a certain subset of positions. Unfortunately, not every mentor will advertise all requirements, so you might apply for a position that you won't even be considered for.

3. **It's often the most professional, most polished student who wins the interview and position**. An email is an excellent way to secure a research interview—if you know how to write it to get noticed. An interview presents the opportunity to demonstrate genuine interest in the lab and an advanced level of professionalism. However, certain "red flags" on an application or at an interview can eliminate a student from consideration even if they would have been an ideal addition to the lab team.

4. **It's a matter of timing. A qualified and professional student who applies for a research position at the "right" time can lock down a position.** The problem is that you won't know what the right time is for a particular mentor. One mentor might screen at the start of the semester, another at the end, and some only review research inquiries that arrive four to six weeks into the semester. (Holiday breaks aren't a popular screening time for most.) The solution is to stand out so you don't have to rely on the luck of applying at the "right" time.

Unfortunately, you can't change the fact that research positions are competitive. **However, there are strategies to be the most competitive**

applicant possible. The rest of this book covers those strategies, including the most common mistakes students make so you can avoid making them yourself.

CHAPTER 5

Your Search Strategy

Ten Search Mistakes to Avoid

A successful search for a research position is about more than reading project descriptions and deciding whether or not to apply. It's also about determining how much time you actually have for a research experience and keeping your search simple so you don't become overwhelmed during the process. Equally important is to know the most common mistakes others make so you can avoid them.

1. **Not making the search a priority.** It's tempting to put off the search for a research position, but doing so won't make it any easier to get it done. As soon as you know you want to participate in undergraduate research, make your search a priority. **Be aware that it can take significant time to identify a research project compatible with your goals in a lab that has an available position**. Start your search early in your college career, and ideally no later than the middle of the semester before you want to begin research. You might be fortunate to find a lab quickly, or you might need to spend half of a semester searching. If it's already late in the semester and you want to start your research experience shortly, don't worry, but start your search as soon as possible.

2. **Placing too much emphasis on reading a scientific paper.** Some undergraduates are advised to read scientific papers as the first step in their search for a research position. Unfortunately, the pressure to do so can derail a search before much progress is made. Attempting to read one of the PI's papers works for some undergraduates, but it's overwhelming for most. For some disciplines, a student will have no hope of understanding the title of a paper much less the methods, results, or conclusion. **It's advantageous to try to read a paper, but if it's too frustrating, don't let it become a reason to delay your search.** (See the next section, Should you read a scientific paper? for more details.)

3. **Not learning something specific about a research program.** Your ultimate goal should be to find a research experience that will become a meaningful and rewarding use of your time. Many students, after struggling to read a scientific paper, become overwhelmed and give up trying to learn about a PI's research program. Whether you learn from a PI's research interests, scientific poster, or advertisement for a research position, **to be the most competitive, and ensure you have a genuine interest in the research, you'll need to learn something about each specific project or lab's research focus before you apply.**

4. **Not scheduling time commitments. It's imperative that you determine how much time you have (and are willing) to dedicate to research before you make a commitment to a project.** When you prioritize the activities that matter to you the most, you exponentially increase your chance for success in research (and elsewhere in college). Do this at the start of your search, so it's easier to pursue or eliminate research opportunities based on your availability as opposed to wishful thinking.

5. **Relying only on advertisements to find a research experience.** Advertisements for research positions are useful when they work. But they can be outdated or filled within a few days of appearing. In addition, not all PIs with available projects advertise for undergraduate researchers. Your search will likely include reviewing advertisements, but might require a more creative approach as well.

6. **Not looking beyond the buzz words.** Buzz words should inspire you to consider a research opportunity, but not be used as the sole factor to *determine* if the position is perfect for you. Your research

experience will be much more than the title of your research project. The opportunities offered are equally important, especially if any of your goals include professional or personal development.

7. **Allowing the possibilities to overwhelm.** Considering every lab, every posted position, every discipline, every research project, and every PI's research interests will lead to frustration. **If you always think the "grass is greener," then you'll never get started on a research project, or you won't stick around in a lab long enough to accomplish much.** Instead, it's best to consider each research opportunity individually and periodically evaluate if the grass is green enough once you've accepted a position.

8. **Searching for a one-size-fits-all research experience.** Even if you accept a position in the same lab as a friend, you won't have an identical research experience. Find inspiration from your classmates and friends, but remember that your research experience will be guided by your goals, the effort you invest in the opportunity, and your specific project.

9. **Pursuing an unobtainable position.** If a research position requires a candidate to have specific qualifications you don't have, then it's unlikely you'll be offered an interview. **Pursuing a position you aren't eligible for will waste *your* time—time you could use to pursue a position you are eligible for.** If the position *prefers* certain qualifications, or states that the *ideal* candidate will have specific qualifications, then it's worth a shot to apply.

10. **Not following a search strategy. Using a strategy-driven search will save time and will help you find opportunities compatible with both your personal and academic goals.** This is a much better approach than randomly picking a lab and hoping it works out or accepting any offer just to be done with the search. A search strategy doesn't have to be complicated to be effective—it's a matter of breaking up the task into small, achievable steps. This chapter is about that process.

Getting Started Is the Hardest Part

For most students, the mere thought of searching for a research position is enough of a reason to put it off. There are several reasons for this failure to launch. First, the search feels like a huge chore. Second, it can be overwhelming, as the term "research experience" encompasses a seemingly endless number of labs and projects, and it's a challenge to decide which to pursue. Third, without having an organized search strategy, the process can quickly become a lot of work for little gain. Add in the standard advice to read a scientific paper before contacting a PI, and it's difficult to remain enthusiastic and find the time to prioritize a search.

However, it would be a shame if any of these reasons discouraged you from starting your search. This is especially true because your research experience could be a defining aspect of your time as an undergraduate.

The purpose of the search strategy detailed next is to accomplish two key search steps in the least amount of time possible. These steps are (1) determine how much time you can dedicate to research; and (2) identify potential research opportunities that inspire you. This strategy is designed not only to save you time but also to be a building block of your research experience. For instance, examining your schedule will help you find a research experience, as well as maintain your academic/life balance once you join a lab.

If you follow the search strategy described in this chapter, the hardest part of your search will be finding the self-discipline to get it done. *You can do that.*

Step 1: Schedule and Prioritize Your Time

Finding the right academic/life balance is tough, and it's harder to do than anyone imagines it will be. But the ability to not only plan your academic schedule but also make it coexist with your "college experience" is essential to both your happiness and your success. To do this well, you'll need a solid time-management plan, self-discipline, the ability to prioritize, and the courage to make some tough decisions.

Part A: Create your schedule in the Google Calendar app

Your goal for this part is to answer the question, "What do I do all day?" To do this most efficiently, use a web-based calendar app. We require our undergraduates to use the Google Calendar app because it is free, feature-rich, and easy to use.[1] With the Google Calendar app, you can schedule all of your time commitments (classes, exams, volunteer activities, social events, etc.), color-code those activities, and display them in a variety of formats. You can view your calendar across most platforms

[1] Full disclosure: We have no financial or competing interests in connection with Google, Cold Turkey, Self Control, or any of the app companies we recommend or mention in this book. Obviously, there are other apps to choose from if you prefer.

(phone, tablet, computer, phablet), schedule new activities, make changes to existing ones, and set reminders so you don't forget the important ones. It's also easy to move your activities to different time slots and ensure that they don't overlap, which is invaluable as you prioritize and make adjustments to your schedule.

Create your schedule. To start, you'll create your current schedule in Google Calendar. **For this to work, you need to be completely honest with yourself about how you currently spend your time — not with how you *should* spend your time.** To stay focused, consider using an app that temporarily blocks your access to distracting websites. We recommend Self Control for Mac users and Cold Turkey for PC users. Then grab a latte and start your favorite music playlist.

Pick a "typical" week in the semester to get started. This step doesn't need to be perfect, so don't overthink it or get stressed about where your time goes. Estimates are fine. You'll make adjustments as needed in part B.

Assign a different a color to each of the following time commitments as you add them to your calendar:

- *Class schedule* (in class and commute time)
- *Study time* (include time spent working on assignments)
- *Extracurricular activities* (clubs, volunteer work, doctor shadowing)
- *Work or internship schedule*
- *Sleep time* (Yes, you need to add this. You're tracking *where all of your time goes,* and sleep is part of it.)

And everything left after that is...

- *Personal time* (social activities, TV watching, exercise, video games, and all the mundane life chores like eating and laundry)

Part B: Evaluate your schedule and make room for research

The standard approach used to determine how much time to devote to a research experience is to ask how much time is *required,* then try to make it work. Unfortunately, this approach relies heavily on wishful thinking instead of solid planning. For some, it leads to overcommitment and, for others, an immediate rejection of a perfect research opportunity because the idea of "X" hours per week feels overwhelming. **A better approach is to determine the amount of time you can dedicate to a research experience and then explore opportunities compatible with that time commitment.** In some cases, you'll be able to consider or rule out a position based on your available hours, but you won't end up over-committed or mistakenly rule out an opportunity because it "sounds" like too many hours.

To determine your ideal lab schedule, first return to Why Choose Research? (chapter 2). Using the highlighted sections as a guide, determine the weekly time commitment you'll need to accomplish your professional research goals. Call this commitment "research hours" and add it to your online calendar in a new color. Ideally, distribute your hours among at least three days per week Monday–Friday, and in three- to five-hour time blocks. In the future, you'll consult with your mentor about your official schedule, but for now you need a placeholder.

If it's not immediately obvious how to incorporate research hours into your schedule, rearrange the flexible areas of your schedule to make room. Determine where you can make changes, where you can move or reduce hours, and which activities you can remove altogether to find time for research. Evaluating your priorities, and adjusting your schedule to reflect them, is an important step in personal development. Knowing what you don't have time for is as equally important as knowing what you do have time for. This is how you learn to organize one of your most important, and limited, resources—your time.

To make room for research, consider the following about your current time commitments and make adjustments where you can.

Personal time Can I reduce the amount of time I watch TV? Can I prioritize hanging out with my friends on the weekends or at night to make space for research during the day? Am I getting the most out of the leisure time I have, or could I give some of it up to pursue a meaningful research experience?

Extracurricular activities Can I schedule volunteer activities to a different day or time or reduce the number of hours I participate each week? Am I getting enough out of each activity to justify continued involvement, or would I be willing to give up something to free up time for research? Am I involved in activities that are important to me, or do I participate in them because it's important to someone else? Can I combine any extracurricular activities and personal time?

Class schedule Can I drop/add a different section at an earlier/later time? Can I take any of my classes online?

Sleep time What is the number of hours I need per night to excel? If I get more sleep at night, could I reduce the number of naps I take during the week? If I increase my sleep time even a small amount, will my study time be more effective?

Study time Can I study more hours on the weekend? Do I have time in between classes that I could use to study or work on assignments? Are the classmates I study with committed to the study sessions, or does the "study" time turn into personal time? If I commit to a study schedule without allowing interruptions, will I be more productive overall? Could I benefit from Self Control or Cold Turkey to make my study time more productive?

As you reflect on the research hours now on your schedule, how do you feel? Imagine that you know your research experience will be a meaningful and rewarding use of your time. Will you be able to meet the time commitment? If yes, then you have a good starting number. **If you feel that the sacrifices you'll need to make are too much, go back through your schedule until you determine a number that will work.**

When done, set preferences to calendar view (not agenda view), and your calendar should look similar to Figure 1.

Additional notes about the schedule shown in Figure 1. The calendar shown is from one of our former undergraduates, MT. It shows three days from the fall of MT's junior year when she had the most activities of her college career.

- To compensate for the busier weekdays, MT made Saturday and Sunday as recovery days with extra personal time. That way, her research could be a priority during weekdays when her mentor was available for guidance.
- To get the most out of her study time, MT used a three-pronged strategy: she only studied with friends dedicated to their academics; she did most of her studying at the library; and she used the Self Control app to limit online distractions.
- Blank spaces are commuting times. MT used that time to grab a snack or latte, check her social media accounts, or chat with her friends or parents.
- MT's schedule included fifteen credit hours of courses in total.
- **Final note: MT was accepted to Oxford University (UK) and medical school (USA) in her senior year of college.**

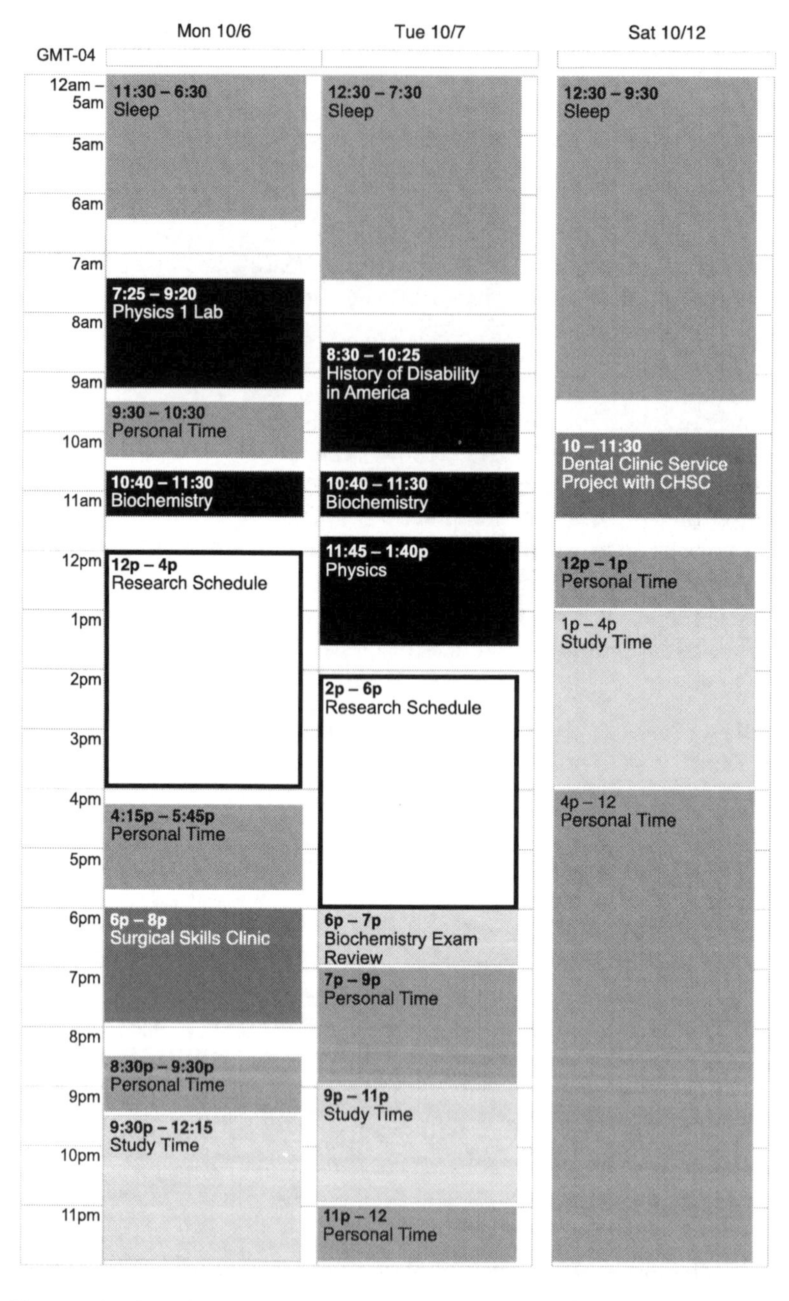

Figure 1. Student class and activity schedule displayed in calendar view for three days of the week. Classes are shown in black; extracurricular activities such as volunteering are shown in dark gray; sleep time and personal time are shown in medium gray; study time is shown in light gray; and research time is shown in white.

Part C: Determine the maximum number of hours per week that you can give to research

The last step is to determine the maximum hours per week you are able (and willing) to commit to a research experience with your current schedule. **Knowing your absolute maximum commitment level will prevent overcommitment because you'll be less likely to pursue a research position that requires more hours than you realistically have available.** It will also be important to know this number at an interview when the required number of hours per week is discussed.

Step 2: Search the Internet

An Internet search is the most efficient way to search for a research position. However, to get the most out of your efforts, in the least amount of time possible, you'll want to do a little preparation before you start.

Download an app

As you identify potential research opportunities, you'll want to collect, organize, and access them quickly. The easiest way to do this is to use an app to save your search results in the cloud so you are able to access the information from all of your devices (computer, phone, tablet, or other mobile device). You'll also want to add new information easily whether it's text you clip from the Internet or a photo you snap on your phone. If you don't already use an app for similar purposes, consider Evernote, OneNote, or Google Docs.

Focus your search

For the most effective search, you'll need to strike a balance between using buzz words for inspiration and considering every possible research opportunity. To do this, avoid trying to distinguish between complex research descriptions such as those for labs that study the neurogenetics of behavior in *C. elegans* or the epigenetic modifications in cancer cells. The fine details will only distract and make your search more difficult. Instead, ask yourself if one (or both) of the research programs are interesting enough to pursue a research position. To help you with this, review the section Basic vs. Applied Research in chapter 3 before moving forward.

If you join a research project that is interesting to you, it will make the difficult parts easier to get through and your overall experience more rewarding. The fastest way to determine if a research program or topic interests you is to focus on how you feel when you read about it. **At least one of three things should be true before you consider a research opportunity. They are as follows:**

- *You like the discipline.* Examples: genetics, microbiology, physiology, cell biology, biochemistry, neurobiology.
- *The lab uses the same or similar techniques that you enjoyed in a lab course.* Examples: DNA isolation, transformation, enzymatic reaction, agarose gel electrophoresis, PCR, molecular techniques, genetics techniques, microbiological techniques.
- *The PI's research interests or an available project just sounds interesting.*

The intimidation factor

No matter what you read, whether it's an advertisement for a research position, a PI's research interests, or a scientific paper, do not become discouraged or intimidated by the scientific jargon. At this point, you are only using these items as a starting point to explore your interests and consider the possibility of applying to one or more of these positions. Here, your only goal is to decide if the overall program, discipline, or techniques are interesting to you. You'll create a short list of positions later.

Customize your internet search

A web search is fast and will provide seemingly endless research opportunities to consider. However, that's also how a web search can derail your self-discipline. The easy part is getting search results—the hard part can be sorting through the massive number that are returned in 0.5 seconds to decide which ones are perfect for you.

The strategy to manage your web search is to use up to five customized search phrases and spend a modest amount of time sorting through the results. You probably won't need to use all five—but if one doesn't give you the results you desire, move to the next. This approach should take you less than an hour to identify about ten to fifteen potential opportunities that look interesting to you. Keeping your search simple is the key to keeping it efficient.

Customize the search phrases offered here to reflect the name of the college or university you attend (and omit the quotes before using them in a search). For example, if you attend the University of Minnesota, major in genetics, and are interested in drug discovery research, you might do a web search with the terms "University of Minnesota undergraduate research opportunities" or "University of Minnesota genetics undergraduate research" or "University of Minnesota drug discovery research," omitting the quotes each time. In many cases, you'll find more than enough opportunities to explore with one to three of the search terms.

Web search phrases

- "Your university" undergraduate research opportunities
- "Your university" undergraduate research scholarships
- "Your university" "your home department" undergraduate research opportunities
- "Your university" "your major" undergraduate research opportunities
- "Your university" research topic[2]

Evaluating the Results

The suggested search terms will produce

- Advertisements for undergraduate research positions
- Links to campus undergraduate research centers or other resources such as college-wide research opportunities
- PI web pages and links to their research interests
- Departments, research institutes, colleges, or centers related to the topic of your search

What you do with the search results depends on what results you get.

Advertisements for undergraduate research opportunities

Advertisements for undergraduate research opportunities can be a gold mine of information. Typically, advertisements for researcher positions cover basic requirements (time commitment, prerequisite coursework, academic standing, additional qualifications), the research topic, the background or significance of the research, and a synopsis about the PI's research program. Advertisements for dishwashing positions, paid or unpaid, typically list the number of hours per week, specific responsibilities of the position, required qualifications, and if the position holds the possibility of "promotion" to a researcher position.

When you read an advertisement ask yourself—

- Does the opportunity seem like something I want to do?
- Can I fulfill the time commitment? (if given)
- Do I have the required qualifications? (if any)

[2]Research topics could include a specific disease (Alzheimer's disease), a research organism (*Drosophila*), a discipline (molecular genetics), or your own personal research-related interests.

If your answer to all three of these questions is yes, then save the advertisement in your cloud-based, note-taking app. Afterward, go through your notes and highlight the name of the project and a sentence or two that covers the main reason(s) you're interested in the position. Even if you are interested in everything the position description describes, only highlight the three sentences (maximum) that you connect with the most. You'll want to find this information quickly in the next step of your search. If your search returned a compilation of advertisements, continue through the lists until you've identified, uploaded, and highlighted sentences in ten possibilities, or until you've reached the end of the list—whichever comes first.

Results that aren't advertisements

Whether your search turns up a department, institute, center, college, or a faculty web page, you'll use the same strategy—glance through the faculty research interests to find inspiration.

What are faculty research interests? **A PI's research interests**, sometimes called research overview, or faculty research description, **is a summary of her research program, problems her research addresses, and the significance of her research**. Some research interests include an overview of techniques used in a research program (such as PCR, mass spectrometry, bioinformatics, cell culture) or broad disciplines related to the research (such as microbiology, cell biology, pharmacology). Many PI's have multidisciplinary research programs, so it's not unusual to see several research topics listed by a single PI.

How to find and use the PI's research interests. To find research interests that are inspiring to you and determine if the work the lab does could be a good fit for your goals, follow these steps:

1. Visit the home page of the department, institute, etc.
2. Navigate to the faculty directory page.
3. Browse the page. When a PI's research interests, program description, or associated keys words sound interesting, copy and save it in the cloud.
4. Highlight a sentence or two that covers the main reason(s) you are interested in the research. Even if you're interested in everything you read, only highlight the three sentences (maximum) that are the most important to you.

5. Repeat these steps until you have a total of ten to fifteen possible research opportunities.

6. Next, visit the personal web page of each faculty member to determine which ones list undergraduates as lab members, advertise projects for undergraduates, or highlight an undergraduate's accomplishment. Consider prioritizing these PIs above those who do not specifically mention undergraduates in their research programs. But, keep in mind that not all PIs who train undergraduates will mention this on their web page—so don't eliminate a lab that inspires you based solely on this information.

Why not all PIs who train undergraduates advertise available positions. There are several reasons that a PI might not advertise for undergraduate researchers. Some PIs believe that the most self-reliant students will seek out a research opportunity without a posted advertisement. Some PIs don't advertise because they recruit from their classes (lab or lecture), or their graduate students recruit from lab courses, or for no particular reason. Of course, some PIs don't advertise because they don't mentor undergraduate researchers. Unfortunately, there is no way to know which, if any, of these apply to a particular PI unless you have inside information.

Other Creative Ways to Identify Undergraduate Research Opportunities

There are several reasons you might want to use other approaches to search for a research position instead of (or in addition to) the web search described in the last section. For example, advertisements for research positions are often outdated. Research positions can be filled quickly, and an ad might remain posted long after the position has been filled.

In addition, not all labs describe their current research on their websites, and not all PIs update their research interests regularly. When you explore research opportunities offline, you might find an inspirational project described on a poster or hear about one at a seminar.

Also, you might interact with a researcher in person, who shares her passion about her project or lab, and you could possibility turn that interaction into an interview.

And finally, most of the approaches described next make it easy to demonstrate that you're creative, ambitious, genuinely interested, or willing to put extra effort into finding a research position. There are no drawbacks to this.

Bulletin boards

Even in this digital world, announcements for undergraduate research opportunities can be found on bulletin boards. Once you know what department(s) or discipline(s) you are interested in, glance at bulletin boards when you're between classes for notices related to research opportunities, volunteer and paid lab positions, seminar notices, and research symposia. If you see an interesting announcement, take a photo of the flyer, and upload it directly to your cloud app. Later, when you have the time, use it as a starting point to explore an opportunity.

Research posters

Research posters are an incredibly valuable resource, yet are often overlooked as a way to search for a research position. **Researchers create posters for professional meetings or symposia and often display them in the hallways outside their labs.** Posters don't tell the entire story of a research project—and that's one of the reasons they are so useful. Posters are a synopsis of a research project or one particular aspect of it. Therefore, compared to a published paper, they are easier to read and understand, but still have enough information to inspire, and provide insight about the research objectives of a project. Although formats vary, a poster may or may not include the background information or significance of the research project, an abstract, techniques used, or plans for the next step of the project. What all posters do contain, however, is a list of contributors to the work done and their affiliations. (That's contact information for you to use!) Spend a few minutes reading a poster to determine if the research appeals to you. If it does, write down three specific reasons, along with the names and contact information of the first and last authors listed. (Note: don't take a photo if you don't have permission.) Then, upload the contact names and poster title to the cloud. You'll learn how to use this information in the next chapter.

Attend a research symposium on campus

Symposia are ideal opportunities to learn about research, meet a potential mentor, and potentially interview for a research position on the spot. Unfortunately, few undergraduates who aren't already involved in research know this, so they miss out on opportunities to attend.

A research symposium might be organized by a department, research institute, student research group, or other campus organization. Although each organizer determines the format of their symposium, you'll likely find a mixture of short research talks (sometimes given by students, postdocs, or a keynote speaker) and an abundance of research posters to read.

If the title of a research poster catches your eye, take a few minutes to read the abstract or introductory paragraph. If you're still interested, continue reading, listen to the conversation the presenter has with others, or (possibly) ask a question to the presenter. **If the poster presenter is an undergraduate, ask, "What is the overall question you're trying to answer?" If it's still interesting to you, ask for the name and title of his mentor (and ideally email address) and who his PI**[3] is. If you want, you can also ask general questions about his research experience that relate to the lab culture and his mentor's training approach. Remember to thank the person (after all, you might become one of his labmates) and ask to photograph his poster before moving on. If the undergraduate is working on something that sounds interesting, you'll contact his mentor by email. You'll learn more about what to do with the opinions of another undergraduate involved in research later in this chapter.

If the poster presenter is a graduate student or member of the professional research staff, wait to ask questions until it's convenient for her—even if you need to return to her poster later in the session. When you have the opportunity to ask about the research project, choose a broad question to get started. **Good questions might be: "What is the big question you're trying to answer with your research project?" or "What is the specific question you're trying to answer right now?" followed by, "Why is that important?" Take notes on her answers because you will use them if you contact her PI to ask for a research position.** You don't have to be perfect at this moment—you just need to show an interest and interact with the presenter as your best professional self.

AFTER you've listened to her research overview, if you're still interested in joining the lab, mention that you're an undergraduate in search of a research project and that you attended the symposium specifically to learn about the research on your campus. **State that you're interested in her project because ________ (fill in the blank with something specific that she told you, or the reason the poster originally caught your eye). Next, ask if she will consider you for an undergraduate research position and tell her that you have X hours per week to dedicate to research.** Yes, this can feel awkward, but you don't want to hint—you want to politely, directly ask for a research position with the most confidence you

[3]A note about posters and author names: more than one PI can be listed as an author on a single poster. You'll want to know which PI the presenter works with if multiple are listed. Also, although typically the first author listed on the poster will be the person presenting, sometimes a different author steps in. You'll need to know the correct contact person when you email a PI or potential mentor or follow up with a presenter through email.

can muster. The ability to do this, no matter how uncomfortable it feels, will only be to your advantage.

The next move is the presenter's. She might (1) conduct a mini-interview with you on the spot; (2) set up an interview appointment; (3) request that you email her; (4) refer you to her PI; (5) state that she doesn't have the time to train an undergraduate researcher and refer you to someone else in her lab; (6) set up a time for you to come and observe her in the lab; or (7) state that her lab doesn't train undergraduates or does not have an open position right now. If the presenter's response is any of the responses 1–6, you've identified a potential lab and possibly a research mentor. Before you move to the next poster, even if she gave reason 5, confirm that the presenter is the first author listed on the poster and which author is her PI. Also, ask to photograph her poster so you can learn more later. Most presenters will oblige, but some won't. Be polite regardless. If you have permission to take a photo, ensure that the entire poster is in the photo and upload to the cloud. Include a note with two or three reasons the research is interesting to you to save you time later.

If the presenter's response was any of the responses 1–5, follow up with her, or her mentor or PI by email within twenty-four hours. Do not wait any longer, because you want to take full advantage of the groundwork you laid by attending the symposium. The sooner you follow up, the more enthusiasm and genuine interest you demonstrate. (Read Step 2: First Contact in the next chapter for how to make the best impression when you follow up.)

Nine tips for attending a poster session at a symposium.

1. **Register when required.** Some symposia have a registration deadline for those presenting posters, giving a short talk, or planning to attend a reception, even if there is no fee to attend. Also, registering might give you access to all the poster abstracts, which can help you decide ahead of time which posters to prioritize. Not all symposia will have printed abstract books, but many make them available online. Some symposia do not require registration to attend, in which case you can just show up and learn.

2. **Start with realistic expectations.** Even at a poster session with an expert explaining the project or overall research program, some information will be tough to understand. You might not have the scientific background to gain a thorough understanding at a poster session. In addition, everyone who is an expert in their project isn't

necessarily an expert at explaining it. Keep this in mind if you feel overwhelmed while listening to a few poster presenters.

3. **Phone etiquette matters.** Put your phone on vibrate—not airplane mode—and keep it in your pocket or backpack except when you take photos or notes. You'll want to upload photos of interesting posters (with permission) to the cloud. You'll also want to upload notes on why a poster is interesting or why the research inspires you. These notes will be important later.

4. **Ask before you take photos.** *For a variety of reasons, some presenters won't want the prepublication data on their poster photographed.* If a poster is unattended, as odd as it might seem, it's acceptable to write down the title and poster authors and some information about the research. Save this information in the cloud so you have it later. Also include a short note with two or three reasons why the research inspires you. When you contact the PI, you'll need to reference the poster title and mention why you think the research is interesting.

5. **Quality photos matter.** If you are given permission to take a photo of a poster, make sure the entire poster is in focus. You'll use the photo later to learn more about the project or review it before an interview. You'll also need the contact information to follow up with the lead author or the PI.

6. **Don't waste a presenter's time.** Not all posters will be interesting to you. It's acceptable to read the title, some text, and move on if you're not interested in learning more. A polite smile and nod to the presenter is all that is required.

7. **Take notes.** *Jot down a few notes on why the poster is interesting that you can refer to when you email the lead author or the PI about a possible research position.* If it's not possible to take notes while the presenter is talking, definitely do so prior to looking at the next poster, because posters tend to blend together.

8. **Be courteous to others.** Venues can be crowded, and sometimes posters and presenters are packed into a very small space. If you want to take a break from the poster session to check your phone, text a friend, or have a social conversation with someone, make sure to step far enough away from the posters that you don't block other attendees' access.

9. **Don't be offended.** Presenting can be tiring, so don't be offended if a presenter isn't interested in discussing his poster with you — simply thank him and move on to the next interesting poster. Although it's possibly rudeness on the part of the presenter, it's not a value judgment against you.

Attend a seminar that sounds interesting

Departments, research groups, and research institutes often host seminars throughout the semester. Some do so as part of a regular seminar series with weekly, biweekly, or monthly talks, and others as an occasional specialty series. The seminar speaker might be a PI from your home institution (or one visiting from elsewhere), a postdoc speaking about his latest results, or a graduate student defending her thesis. How to follow up with the speaker (or the host) depends on who the speaker is and her affiliation. **Attending a seminar can inspire you to work on a particular project or research topic and help you to both identify and impress a potential mentor.** Seminars are generally open to everyone. So if you hear about one from a professor or see one advertised on a bulletin board, a departmental web page, or social media page, or receive a notification about one from a listserv, you're invited to attend. **Ideally, before you attend the seminar, spend ten to twenty minutes on the Internet reading the research interests of the speaker to help you understand the seminar and familiarize yourself with the research topic.** At the very least, do a web search using the entire title of the seminar as the search term.

Although a doctoral or master's degree defense is often open to the public, and you could attend, we don't recommend it to find a mentor. A graduate student who is defending a thesis will not be available to mentor a new undergraduate researcher.

What to do when you arrive at a seminar. **Look for the posted seminar notice. Take a photo and upload it to the cloud. Make sure you have the talk title, date, name of the speaker, and the host's name (if applicable).** Even if you previously photographed a notice on a bulletin board, do it again because sometimes the talk title changes. You need to be accurate with the seminar title when you email a potential PI, or you'll lose credibility. If there isn't a seminar notice posted outside the room, write down the date, speaker name, and title of the talk — you'll get this in the first few minutes of the talk and during the introduction. Also, write down the name and department affiliation of the person who introduces the speaker. **If the speaker is from your campus, you'll contact him for a re-**

search position. If the speaker is from another campus, you'll contact the person who introduces him to ask for a research position.

You have one job: leave with what you came for. **During the seminar, make it your goal to write down three reasons, phrases, or sentences that highlight why the research is interesting to you.** It could be a technique, a result, or a statement about the project background or significance. The more specific your notes are, the bigger the advantage you'll have when you apply for a research position.

Practice safe seminar. When you attend a seminar, you'll want to elevate yourself to the same professional level as the other attendees (or above, if you sit by someone who is rude). A seminar room is filled with busy professionals who believe the speaker has something to contribute either to their own research or to the world at large. Often, a seminar might be the only chance the researchers in the audience will have to hear the person speak. Most attendees will be professors, professional research staff, and graduate students. **Although undergraduates are welcome to attend seminars, unless a professor is awarding extra credit for doing so, it is uncommon. Therefore, undergraduates who do attend will be noticed.**

We'll share a story that underscores the importance of this very point. One of our colleagues occasionally forwards seminar notices to students who express an interest to join his lab. He finds that almost no students attend, but those who do either become inspired by the research topic or realize that it's not something they wish to pursue. One semester, he informed several students through email about a seminar closely related to his research program. On the day of the seminar, although the room was large, it was packed with some attendees standing in the back or sitting on the auditorium stairs. Approximately fifteen minutes into the seminar, an undergraduate student a few rows away from our colleague proceeded to remove several items from his backpack and place them on the desk: a binder of class notes, a highlighter, a pen, a bag of candy, and hand sanitizer. The student then began what our colleague presumed was the student's typical routine in a class lecture: he applied the hand sanitizer and spent the rest of the lecture highlighting notes in his binder while pausing occasionally for a very crunchy treat from his snack selection. On occasion, the student would stop to check his phone, presumably to send a text or answer email, and glance up at the speaker in the front of the room.

Beyond this ritual creating a distraction for our colleague and everyone else around him, the student took a seat that could have been occupied by someone who was interested in the seminar. Imagine how short the research interview was when that same student showed up a few days later at our colleague's office in hopes of securing a research position. The student had no hope of convincing our colleague that he was interested in research or that he had the ability to focus on research for fifty minutes at a time. Although this is an extreme example, even if the student had spent the seminar distracted by his phone instead of the panoply before him, the interview would have been just as short. As an undergraduate, your attendance will stand out—it's up to you to make it an advantage in your search.

Four tips on seminar etiquette that can make all the difference.

1. **Sit in the back in case it's difficult to focus (or stay awake).** If needed, distract yourself by taking notes on a legal pad to keep yourself from doing the falling-asleep head bob.

2. **Turn off your cell phone** (or put on airplane mode). It's bad form to text, check your email, or otherwise broadcast that you have something more important to do than focus on the speaker. In a professional seminar, this matters. *Even if everyone else around you is live tweeting the seminar, stay off your phone.* As an undergraduate, what you do will be perceived differently than what a professional researcher does. It's not fair, but it's true.

3. **Take notes on paper.** Again, use the legal pad. Some believe that those who take notes on a computer, tablet, or cell phone are actually on social media or playing games. Take notes on paper to avoid this misunderstanding.

4. **Don't leave the seminar early.** Most seminars last around fifty minutes—you can get through it. You can respectfully leave when, at the end of the seminar, the speaker invites questions. However, you might learn something from the question-and-answer session if you stay.

A note on why it's a bad idea to claim you attended a seminar if you did not. Seminars get canceled. They get rescheduled. A talk title can change at the last minute, and so can the scheduled speaker. Sometimes seminar rooms are small or only a handful of people attend. It's not worth the risk to falsely claim attendance at a seminar that you did not attend.

You'll risk eliminating yourself from a position due to a "character issue," and once a PI has decided to reject a student because of one, there is virtually no way to be reconsidered.

Ask your professors if they have space in their lab

(Note: Tips on how to approach a professor in person, or through email, are in chapter 6, Your Application Strategy.)

In the sciences, many professors direct research programs in addition to teaching classes. Typically, classes overlap with a research specialty, so if you're interested in the subject a professor teaches, you might be interested in her research program. Some professors even recruit promising students from the classes or labs they teach. **You have the potential to be recognized as a promising student in a lecture or lab class, so use it to your full advantage.**

You already know what you need to do:

Do your best work. To some, how much effort you put into a lecture class is indicative of how much effort you'll put into research. Because professors often teach their research specialty, low grades (on a quiz, exam, or extra-credit assignment) could signal disinterest in the subject, and therefore disinterest in their research program. *Some professors will only consider undergraduates who are near the top of each class for a research position.*

Be your best self. This means, in addition to doing your best work, avoid arriving late to lecture or being a distraction to your classmates or the professor by chatting, texting, eating, or sleeping in class. Don't leave lecture early, and stay off your phone and social media during class. Most professors won't call you out in lecture, but can tell when you're engaged in "distracted learning" based on your typing cadence and how often you lift your phone up or tilt your head down. It. Gets. Noticed. If the lecture class uses peer-to-peer learning, contribute on every assignment and always be pleasant to your classmates — even if someone irritates you.

Ask your teaching assistants (and let them recruit you)

If you have a teaching assistant in the sciences, he probably does research. Look up his research interests on the web and determine if the project he works on is interesting to you. If your lab class has several teaching assistants, decide who has the most inspiring research interests. **Approximately three weeks into the lab course, approach your teaching assistant and say, "I read about your research interests online and would**

like to learn more about your project because ________" (fill in the blank with the specific reason that you're interested). Then continue with, "I have X hours per week to dedicate to research. Will you consider me for a research position?"

What you do next depends on his response. If your teaching assistant says that he doesn't have an available position, ask him if he knows of anyone who does. In person, to a teaching assistant, you can be more flexible when talking about your research interests, and it won't count against you. Compared to faculty members, graduate students are more likely to have a conversation about unfilled undergraduate positions. If your teaching assistant doesn't say yes, but doesn't say no, it's not a bad sign. If he's interested in mentoring an undergraduate researcher, he'll likely set aside time to interview you at a later date. Alternatively, he might refer you to someone in his lab or to a listing of available undergraduate research projects. He also might take the "wait and see" approach, which essentially puts you on a semester-long interview. This can be a huge advantage if you perform well (and a disadvantage if you slack off in lab class).

At the end of the semester, or possibly sooner, you might find that the teaching assistant recruits you. **Some of my (DGO) most successful undergraduate researchers were recruited from a lab class by my graduate teaching assistants.** If your teaching assistant doesn't follow up with you within four weeks of your asking about a research position, ask about the possibility of joining his lab one more time. If his response is still noncommittal, direct your search elsewhere.

Seven tips to impress your lab's teaching assistant You'll want to show the teaching assistant that you're smart, that you take your lab responsibilities seriously, and that you get along well with your classmates. To do this, make sure that you do the following:

1. **Be the ideal student.** Show up on time, prepared, and ready to learn. Complete all extra-credit opportunities and put focused attention into your lab assignments. Turn assignments in on time and make sure to follow all instructions for lab reports and papers.

2. **Demonstrate perseverance.** If an experiment is hard, frustrating, or fails, keep a positive attitude and do the best you can. Avoid giving in to frustration or complaining. Instead, see the next point.

3. **Demonstrate that you want to solve problems.** If you, or your lab group, makes an error, try to approach your teaching assistant with

the question: "What can I/we do to fix this mistake?" or with an idea to solve the problem.

4. **Participate but don't dominate.** If there are group assignments, always step up to contribute your fair share to the group effort, lead when possible, but don't try to control the other members. Step back when needed to show you can work well as a member of a group as well as take the lead.

5. **Be patient with others.** Be both understanding and patient with your fellow group members if someone makes a mistake—even if he ruins the experiment. Remind yourself that although it might be annoying, it's not tragic. Your response to your classmates' mistakes will indicate how easy (or difficult) it will be to work with you in a research lab.

6. **Demonstrate self-reliance.** Ask for help when you need it, but double-check that the answer isn't already in a protocol, lab manual, or the syllabus before you ask. You'll make a negative impression if you ask your teaching assistant to essentially read a protocol or instructions to you.

7. **Treat each interaction as a mini-interview.** It's best to view every interaction you have with the teaching assistant, every question he asks you, as part of his selection process. Some teaching assistants might conduct several mini-interviews during the lab class that seem more like casual conversations.

Contact former teaching assistants

If you've already completed a lab class, do a quick web search for your former teaching assistant. You should find a short statement about her research project—what she does and why it's important—and be able to determine if it is interesting to you. If it is, contact her through email or attend her office hours. **Don't drop by her lab to surprise her without an appointment**. She may or may not be there or may be in the middle of a protocol that she can't walk away from. Essentially, dropping by a lab without an appointment isn't an effective way to secure a research position, but it can be an effective way to annoy other lab members who are trying to work.

Check with student organizations

Start by examining the web pages (social media and official university pages) of student organizations. Consider preprofessional societies, honor societies, and groups for a specific major or career path. **Basically, if the group's members participate in undergraduate research, it's worth exploring.** The group might post advertisements for PIs or mentors searching for undergraduate researchers. These will likely fill quickly, but it's a good option to explore just in case.

If the group hosts speakers from your campus searching for undergraduate researchers, attend these meetings. **Before a talk begins, snap a photo of the slide that has the presenter's name, title, and department or center affiliation.** Whether or not you talk with the speaker, you'll want her contact information so you can email her shortly after her presentation. **During the talk, write down three specific statements or phrases about her research, which are interesting to you.** It can be a technique such as "fluorescent labeling of proteins in living cells" or the central question addressed by the speaker's research, "Our lab studies a family of proteins that inhibit...," or a specific result, "We found that protein A interacts with protein B..."

If you have the opportunity to discuss a potential research experience with the speaker after her talk, remember to approach your conversation as if it's an interview, *because the speaker definitely will.* If you would rather follow up with her by email, do it within a few hours and no later than twenty-four hours. The sooner you follow up, the more you demonstrate your enthusiasm, and more likely you'll beat your competitors to the in-box. The next chapter will tell you how.

Ask your classmates about their research experiences and evaluate their answers

By asking a few questions, you'll be able to determine if your classmates' research experience sounds like something you'd enjoy. **However, remember that whether they like or dislike their research experience isn't as important as the reasons they feel the way they do.** For instance, if they like their research experience because they get plenty of time off to study for exams, or regularly skip lab when there isn't anything to do, would that experience be a good match for you? What if they dislike the experience because they spend most of their time observing or cleaning animal cages? Would you form the same opinion if you had a similar experience? If they feel overwhelmed by the weekly hour commitment, is it because their mentors' expectations are too high, or they haven't found the proper life/academic balance?

It's not enough to ask your classmates about their research experiences; you need to evaluate what they tell you based on your goals and the expectations you have for your research experience. And you need to remember that your classmates will be passionate (or not) about their experience because of what it means to them, what activities they give up to participate in research, and what they want to gain from a research experience. Also keep in mind that they might be frustrated with a difficult technique or experiment, which could negatively influence how they feel about their research experience when you speak with them. Perhaps in a week, they will be through the rough spot and once again be inspired by the benchwork.

Questions to ask your classmates. For the conversation to be the most effective, ask for specific information about their research experience. Here is a list of the most important questions you should ask.

- What do you like the most about your research experience?
- What do you dislike the most?
- What has been the most challenging part of your research experience so far?
- What is the name and title of the person you initially contacted about the position? (If they applied to a specific person, contact that person if you decide to pursue an opportunity in the same lab.)
- How many hours per week do you spend in the lab, and in what blocks of time?
- Do you set your schedule, or it is set by your mentor?
- What techniques have you learned?
- What do you spend most of your time doing? Or what is a typical day in the lab like?

The following question is not essential, but might give you additional insight:

- What are the objectives (or specific aims) of your project?

If your classmate applied in response to an advertisement, ask where it was posted. Check the same place for a current advertisement. If there isn't one, don't be discouraged. If you're interested in the lab, follow up with the person they applied to, or contact their PI.

Check with the campus office of undergraduate research

Your campus office of undergraduate research is an important resource, so don't overlook it. Explore both the official web page and any social web pages to find out about available research positions, programs, scholarships, and events geared toward writing research proposals, abstracts, designing posters, crafting CVs and resumes, and much more. Many offices also have specific information about summer and off-campus (often paid) undergraduate research opportunities. Join their Twitter, Facebook, or Google+ feeds to stay current on the most relevant information.

Ask academic advisors (directors of undergraduate programs and major advisors)

An advisor for your major, or a preprofessional track, might have additional suggestions on where to look for research opportunities. An advisor probably won't have a list of PIs with available positions but might know which professors regularly train undergraduate researchers. Advisors also might know of a specific database that advertises undergraduate research opportunities, know about a departmental or college-wide research symposium, or be able to direct you to specific research fellowships.

Sign up for even more email (really)

Join all email listservs related to your major, preprofessional groups, and clubs that are research-related. **Skim each email the day it arrives**. In particular, pay attention to interesting seminar announcements, research symposia, or available research positions. **If you read an announcement for an undergraduate research position,** know that the professor or mentor will be inundated with responses within a few hours. So you'll want to **respond ASAP, with a custom Impact Statement, and any supplemental materials asked for in the email** (transcript, CV, or other). This is one of those cases where the first few responses might be the only ones read.

Look at PIs' social media pages

Some PIs have social media pages open to nonlab members that are useful to determine if the PI trains undergraduate researchers. Although more of a long shot, it can be worth a quick look to determine if there are any undergraduate research opportunities listed.

Consider a job washing lab dishes (or doing research-related tasks)

When available, ads for these positions are typically found on departmental bulletin boards, in online job databases for undergraduates, or advertised through departmental or club listservs. However, before you accept a position, confirm that an option to be "promoted" to researcher is on the table. Sometimes students are hired for a dishwashing position without the possibility of becoming a researcher. Also, remember that if you receive a paycheck for undergraduate research, you might be ineligible to receive course credit as well, and it could affect your financial aid packet. Consult with your financial aid office and your advisors before you make a final decision.

Explore outside your major and your college

Regardless of your major, consider searching outside your home department for research opportunities. Many research programs use a multidisciplinary approach and train students from various majors. For instance, undergraduates in our lab have majored in physics, biology, molecular biology, microbiology, genetics, bioengineering, psychology, and biochemistry. The majority have been premed (or prehealth), but others have been predental, prevet, pregrad, or planning to enter the job market after graduation. Your goal should be to participate in a research project that excites you, in a lab where you can accomplish your personal, professional, and academic goals. Whether or not the lab is in the same department as your major isn't always a relevant factor.

Explore outside your campus

If your college doesn't offer undergraduate research opportunities, or you want to participate in a specific research program, you should consider a research experience beyond your home campus. **Opportunities include programs at a research company, national lab, research institute, or another university.** Make it a priority to visit your college's center for undergraduate research for advice. Some programs take place over the summer and others during a semester. Many include the opportunity to focus on research without additional academic responsibilities. Some provide a stipend or fellowship, and some provide a salary in addition to room and board.

If you want to pursue one of these options, start your search early because many have application deadlines months before the program start date. Also, you'll need to fulfill the eligibility requirements and complete a formal application for each program. Some require a combination of personal essays and letters of recommendation to complete the application. If you want to apply for a paid position, first do a web search for "paid summer undergraduate research programs" for an extensive list of possibilities. You can also leave "summer" out of your search terms to include programs that take place during the fall or spring semesters. Alternatively, search using the terms "undergraduate research national lab" or "undergraduate research internship," with or without the term "summer." And don't forget about a very basic search in your field of interest such as "undergraduate research chemistry," which, depending on your field, might return numerous off-campus opportunities to explore.

Maintaining Your Academic/Life Balance in All Future Semesters

Every semester is a do-over with respect to scheduling your academic/life balance. Use the scheduling approach presented earlier as a long-term strategy to schedule classes, activities, and research hours for continued success. **Evaluate your schedule often and add, reduce, or eliminate items as needed to keep your activities in check without becoming overextended.** The small effort spent scheduling is worth the big payoff in productivity, stress reduction, and overall happiness. Plus, the self-discipline and time-management skills you'll gain by periodically evaluating your schedule throughout college will lead to unparalleled advantages in both the short and long term.

Why it's important to schedule study time

If you do the majority of your studying a few days before each exam, you'll need a new system as soon as you have a research position. The reason, beyond keeping your academic/life balance intact, is that the more last-minute "stuff" you have to do, the more likely you'll cut your research hours to get it done. On the surface, this might seem like a good solution, and on occasion it might not create an issue. However, **each time you cut your research hours, you send the message that you might be overcommitted, managing your time poorly, or uninterested in your research project.** If reducing your lab time becomes a frequent occurrence, at the worst, it could lead to your dismissal. At best, your mentor might be concerned that you're on the verge of quitting, or you'll earn the reputation of being unreliable, or your opportunities in the lab will be limited.

Therefore, schedule your study time and *find the self-discipline to stick with it.* When you do need extra hours to study or complete an assignment, try to take them from your personal time—not your class, sleep, or research schedule. For more specifics on why cutting your lab time is a risk, see The Time Commitment in chapter 4.

Why you should avoid overcommitment

Finding the academic/life balance is so much easier said than done. After all, there is an entire industry dedicated to time-management strategies aimed at professionals who "should" have it figured out. For you, good time-management includes understanding your priorities and your limits as well as recognizing the signs of overcommitment and adjusting your schedule, if needed.

When you overcommit your time and fall behind in the social or professional obligations you've made, it adds stress to your life—a lot of stress. You end up being stressed about the studying you're not getting done, stressed about letting your friends down, and stressed about the obligations you abandon just to keep your head above water. **All this stress takes a toll on your capacity to learn and recall information.** If it gets bad enough, you'll become unable to focus during lecture or while reading assignments, so you retain less information, and by necessity end up being even more stressed as exams approach. **So inevitably, you try to solve the problem in the short term** by cutting class, getting less sleep, reducing your lab hours, cramming for exams, or doing subpar work on your out-of-class assignments. **This in turn could compromise your academics, recommendation letters, happiness, and potentially your health**. Worse than the feeling of letting a club or a friend down is the personal toll overcommitment takes on you, because you set yourself up for failure instead of success.

Every part of your life is more difficult when you are overcommitted. Make a conscious decision to evaluate your schedule before adding new activities, or quickly reevaluate and prioritize if you start getting in over your head. And be absolutely honest with yourself about where your time goes, or no amount of thinking about your schedule will help. For example, if two hours of scheduled personal time regularly turns into five, or one hour of scheduled study time often turns into fifteen minutes of study time and forty-five minutes of personal time (or nap time), acknowledge it and make changes.

Why you should avoid sleep deprivation

It's understandable to want to sacrifice sleep over anything else to make your schedule fit. And, for some, doing this the first semester or year of college seems to work with no significant consequences. But as your classes become more challenging, you'll need more time to study. At the same time, the amount of information you'll be expected to learn in a research position will also increase. If you consistently choose to forgo sleep, it will lead to sleep deprivation. This matters because you cannot excel when you're sleep deprived—function, yes; excel, no. (Five hours or fewer of sleep per night counts as sleep deprivation.) Your motor skills and abilities to focus, learn, and recall information are all compromised when you are sleep deprived. For the most part, you probably don't mind. Running on a few hours of sleep each night—as long as you're getting the term paper done and are doing well on exams—seems worth it. It probably doesn't even feel as if you're sacrificing anything. It might even seem brag-worthy.

However, although the difference between your performance when you excel and when you simply function may not be a problem to you, it will become one for your research mentor. When you're sleep deprived, you look like it, and it will affect your performance at the research bench. If your mentor believes mistakes or slow progress are due to consistent sleep deprivation[4], regardless of your motivation or enthusiasm for the project, it will have negative effects on your opportunities for advancement, your research responsibilities, and possibly your recommendation letter. Worse case, it could lead to your dismissal from the lab. And, of course, research aside, there is no area of your personal or academic life that will be spared the effects.

One final note on academic/life balance. If it's difficult to maintain your academic/life balance, make an appointment with the campus counseling center or study center. You'll no doubt learn additional strategies to help manage stress and maintain your academic/life balance.

Should You Read a Scientific Paper?

The standard advice given to undergraduates is to identify a potential lab and then read one of the PI's papers. In theory, this is an easy idea. In practice, most students find it overwhelming, and it becomes a reason to put off their search. However, that doesn't mean that you shouldn't try to read a paper. It simply means that if doing so becomes frustrating, you shouldn't let it derail your search. If you choose to read a paper, or if the PI gives one to you as part of the application process, use the following tips as you go through the process.

Ten tips for reading scientific papers

1. **Find a paper.** To find a paper, search PubMed or Google Scholar using the PI's name or visit her website to view an abbreviated list of publications. Choose two or three at most but focus on a single paper at the start. Move to the next paper if you get stuck on the first one.

2. **Don't give in to frustration.** Keep trying even if most of the language or concepts are confusing. You might not have the background knowledge or scientific vocabulary to gain a meaningful

[4] This is not in reference to a day or two here and there, as everyone has those days. We're referring to several times a semester, which leads to a consistently weak performance.

understanding of what you read, but you can still learn something. For papers in some disciplines, you won't understand the title without performing a web search, much less the results, significance, or conclusions. *It's important to remind yourself that difficulty is to be expected and not let it throw off your enthusiasm.*

3. **Set realistic goals.** Your goal isn't to become an expert in the subject or to be able to discuss the paper at length, but to introduce yourself to a topic that is important to the PI and learn what you can. If, while reading a paper, you determine the research program might be a good fit for you, consider it a bonus.

4. **Use resources to decipher the terms.** Use web searches and textbooks to decode the significant terms. Look up short terms, such as "epigenetics," or longer ones such as "laser scanning confocal microscopy." You might find it easier to understand the information this way, even if it is generalized.

5. **Focus on the summary section (or introduction).** Once again, use a web search to glean the overall focus of the paper and decide if it's interesting to you. After you read the first few sentences of the introduction or abstract, you might have an understanding of the general problem that the paper addresses.

6. **Glance at the methods and materials section.** Determine if any techniques are ones you learned about in lecture or performed in an instructional lab class. If so, you'll have a point of reference. Again, do a web search to read about or, when available, watch videos of specific techniques. At this stage, try to learn why the techniques are important to the research topic — not just how they are done.

7. **Make a list of three to five interesting points or facts.** PIs want to nurture genuine excitement and, whenever possible, send a project in the direction that inspires a student the most. If you understood the paper well enough to be inspired, the PI will definitely want to discuss why.

8. **Papers are often collaborative.** A paper might list multiple authors from multiple labs. A technique or experiment you find interesting might not have been done in the lab that you're interested in joining. Some papers denote each author's specific contribution, but many do not. Be aware of this if you plan to contact a PI based on a specific technique or an experiment described in a paper.

9. **Some papers are published on projects that have been completed or abandoned by the PI.** These projects are typically related to the overall research focus of the lab but are no longer being pursued. Even if a specific project is no longer available, if you like the research focus of the lab, it is likely that other interesting projects will be available.

10. **Revisit the paper after you join a lab.** Once you join a lab, you'll have someone (or possibility several people) available to guide you through a paper and help you to understand it. When you find a paper that is interesting, upload it to the cloud so you don't have to search for it when the opportunity to discuss and understand it on a deeper level arrives.

Why it's a bad strategy to claim you've read a paper when you haven't

When you apply for a research position, you'll always be better off being honest about what you read beforehand—even if it's nothing[5]. If you mention that you've read one of the PI's papers, and it helps you get the interview, it's likely that you'll be asked to name the paper and explain what you learned or found noteworthy from it. **If you did not read a paper, it will be obvious within seconds because you can't fake your way through a discussion of a scientific paper**. During interviews, numerous students have said, "I read a paper of yours," and when I (DGO) ask the inevitable follow-up questions, "Which one?" or "What was the most interesting part of the paper?" the answers have always been a flustered, "I don't remember," or a variation on that theme. Typically, the next moment is awkward as the student quickly states things such as maybe it wasn't a paper, or they tried to read a paper, or that they are confused and didn't mean that they read a paper, or that they *planned* to read a paper. In any case, it's disappointing—not that the student didn't read the paper, but that they misrepresented doing so. Colleagues I've spoken with have echoed similar experiences.

[5] I'm (DGO) perfectly fine with the fact that no undergraduate has read one of my papers before joining my lab. As long as a student is honest about not doing so, it's not an issue. I'd rather choose a student who demonstrates solid character and honesty from the start, rather than risk mentoring one willing to misrepresent something as trivial as reading a paper.

CHAPTER 6

Your Application Strategy

Ten Application Mistakes to Avoid

FOR many PIs and mentors, their first impression of an undergraduate is the one they use to determine if they should offer an interview. Therefore, whether in person or through email, when you apply to a research position, it's essential that you use a professional approach at every step. Here are the ten most common mistakes undergraduates make when applying for a research position, tips on how to avoid them, and why you should. In the interest of brevity, PI will be used to refer to either a PI or mentor in this chapter.

1. **Sending an unprofessional email. The email you send matters.** It can be the most powerful, effective method used to secure an interview, or it can immediately ruin your chances for a particular research position. The quality is so important that it has a dedicated section in this chapter.

2. **Not demonstrating specific knowledge about the research program and a genuine desire to learn more about it.** PIs know that if you're genuinely interested in a research project, you'll have a more meaningful and rewarding research experience. You'll also be more successful with research, and all PIs want their undergraduates to be successful. **Unfortunately, many students who would be an ideal**

fit for a lab send the opposite message when they first contact a PI. In particular, common mistakes include these:

- **Sending a generic email.** Anything along the lines of, "I read about your research and thought it was fascinating," or "I saw your research advertisement, and I'm intrigued," is generic and seems inauthentic. Using those phrases is also a lost opportunity to get your email noticed because they are overused. **The best way to avoid a generic email is to include a customized Impact Statement in your email. An Impact Statement is a short statement (one or two sentences) that demonstrates what you've learned about the PI's research program and that shows you're genuinely interested in what you could learn in her lab.** Whether you state the specific reason a project or research program is interesting to you, or what you hope to learn from a specific research position, customization is the key. **Being specific is essential.** It's not enough to say that a research opportunity is fascinating—you need to state *why it's fascinating to you.*
- **Applying to the wrong position**. If you apply to a neurobiology lab, but state a passion for barn owl ecology, or apply to a lab that specializes in cancer drug screens, but state that your dream is to study marine mammal population dynamics, it's clear your interests do not match the PI's research program. **A PI doesn't want you to accept a position if it's not right for you, so most will pass if you have significantly different research interests from theirs.** Plus, you simply shouldn't pursue a research position that doesn't resonate with you. Research can be difficult, it has boring phases, and it costs you time that could be spent doing other things. For it to be worth it, you need to be interested in it.
- **Viewing research positions as interchangeable.** Avoid using any statement similar to, "I'd love to participate in your cutting-edge research program, but if you don't have an available project, can you refer me to someone who might?" Students often include this statement thinking that it demonstrates pure enthusiasm and excitement for research. Unfortunately, it does not and actually sends the message that they believe that all research projects are the same. **PIs choose undergraduates who want a specific opportunity** because they feel a connec-

tion to the research or value what they can learn from a specific project or topic.

- **Asking the PI for an appointment to explain her research program.** On the surface, this seems like a good idea. After all, research programs are complicated, and who better to explain the details than a PI? Therefore, it can be tempting to send an email similar to, "I'm fascinated by what you do, can you tell me more about your research or give me a tour of your lab?" The problem with this approach is that you miss the opportunity to demonstrate both your desire and your ability to learn on your own — two things that most PIs believe are essential. You also risk sending the message that you might be a do-the-minimum kind of student, which isn't a competitive strategy. Obviously, this isn't the message you intend to send, but it's the one that is often heard with this approach. You can avoid this mistake by crafting a customized Impact Statement. You'll learn to do that in the next section.

3. **Emphasizing certain personal development goals.** Every PI knows how a valuable a research experience can be to develop, refine, and demonstrate personal development. However, many will hesitate to offer an interview to a student who emphasizes certain aspects during the application stage. **Especially when applications are done through email, some personal development goals can be misinterpreted.** For example, stating, "I would like to learn some better methods for concentrating when something isn't interesting," sends the message that the undergraduate might not be interested in the research or might be difficult to work with. As you establish a mentoring relationship in the lab, your personal development goals, and how you can achieve them, will likely become part of the conversation. But **at the application stage, most personal development goals are best left unstated unless you're specifically asked about them.**

4. **Not following instructions.** Because so much is riding on an undergraduate's ability to carry out instructions in a lab, **many PIs use the application process as test to screen out "risky" students. In those cases**, with an online or email application process, **an undergraduate who doesn't follow all instructions perfectly won't be considered for a position. Therefore, always take the time to read each advertisement, application, and email thoroughly.**

Give acute attention to the instructions to ensure that an avoidable mistake doesn't prevent you from getting an interview.

5. **Delaying a response.** Once started, it's a mistake to postpone any part of the application process. **If you don't follow up within one day** (or two, maximum) **after being asked for additional information** (such as a transcript or resume or to fill out an application), **you might put yourself out of the running.** Even if you follow up after a week with an explanation for the delay such as, "Sorry, I had exams/was sick/parents were in town/didn't have Internet access." The reason doesn't matter if the PI has already selected other students for interviews. Therefore, respond quickly so you don't risk losing the opportunity to another candidate and to demonstrate that you're serious about the position.

6. **Asking for a salary for an unpaid position.** Some lab positions include a paycheck, but **unless a position is advertised as such, you should presume it is for volunteer or class credit only.** Actually, asking any derivative of, "What kind of compensation can I expect if I choose your lab?" is a risky approach when you first contact a PI. It can send the message that you're interested in a position only if you'll be offered more than the chance to learn, contribute to something bigger than yourself, and have opportunities to build your professional and personal development. It can also send the message that you don't care about the science the lab does. Therefore, unless it's a deal breaker, wait until the interview to ask about the additional perks.

7. **Appearing to be overcommitted.** A student once told me (DGO) that his biggest challenge in research would be to "not prioritize my research project if I get behind on studying." The student intended to demonstrate a dedication to research, and on the surface this might sound like a PI's dream. However, the message he sent was that he anticipated having trouble managing his time. **Most PIs will hesitate to offer a position to a student who appears to be overcommitted or is unsure how they will handle their academic/life/research balance.** There is too much risk involved, and if it doesn't work out, it's not good for the student, and it's not good for lab. (If you haven't considered your schedule yet, return to the Step 1: Schedule and Prioritize Your Time section of chapter 5 as soon as possible so you can answer scheduling questions with confidence and honesty at an interview.) Everyone prefers students who

can balance academic, research, and social obligations from the start of their research experience.

8. **Not having the required qualifications for the position.** Whether it's academic year, prerequisite coursework, experience, or enough research hours available, a requirement is a requirement. If you don't have the qualifications for a specific position, don't apply. **Your time will be better spent applying to a research position that you're eligible for.** Most PIs will leave a position unfilled rather than offer it to a student who doesn't have the essential qualifications.

9. **Misrepresenting anything during the application process.** Before you include something on your application, ask yourself this: "Is it truthful? Is it accurate?" If it's not, don't include it. A PI must be able to trust each person in her lab to work with integrity when recording results and data, when conducting research, and when working with others. If a PI doubts your character or credibility, she probably won't call you on it, but she definitely won't offer you the position either. **The most common misrepresentations on undergraduate research applications are these:**

    - **Easily verifiable information.** A PI can double-check a transcript, GPA, major, and academic year. If certifications, vaccinations, specific training, or a financial aid award is a required qualification, proof will likely be needed prior to the start date and sometimes at the interview.
    - **Work history or volunteer experience.** Even if a PI doesn't check on employment dates, volunteer activities, or contact references prior to an interview, everything listed on a CV is fair game during the interview. It's immediately obvious when items are inflated if a student is unable to answer specific questions about the experiences listed on their CV or overstate their involvement in an activity. These students have short interviews and rarely end up with an offer to join the lab.
    - **Proposed time commitment.** This includes both the hours per week and total number of semesters you plan to commit. It will be awkward in an interview if you're offered the position and then say, "Oh, I can't be here in the summer like the advertisement stated," or "I can't commit to the number of hours I stated on the application." At best, you'll annoy the interviewer, which is not a good start to a research experience.

Even worse, he might immediately stop considering you and won't recommend you to a colleague.

- **Intended career path.** Typically, PIs don't offer a research position because of a student's long-term career goals—after all, those could change over the course of a college career. However, **undergraduates who declare one path, but are obviously preparing for another, could lose out on a research position at the application stage.** A CV and transcript will reflect the intended career path, so there is no point in pretending otherwise. The most common example is a student who declares a pregrad or undecided path when they are obviously premed. If you've heard that some PIs won't train premed students, there is no need to be concerned. In truth, if you're premed, you don't want to end up in one of those labs anyway, because they aren't compatible with *your* goals.
- **Having read a paper.** Some undergraduates feel pressured to claim that they read one of the PI's published journal articles even if they haven't. Don't give in to this pressure. We've already covered why it won't help you in the previous chapter.

10. **Not scheduling time to get the applications done.** For many undergraduates, around the time they are ready to send out applications, the excitement of a research position starts to wane. It's not that they've changed their mind but that the process starts to muck with their enthusiasm. If this happens to you, dig deep to stay self-disciplined to finish it up—don't put off your search. **The strategy in this chapter will help you get through this quickly and help you submit quality applications.**

Application Procedure

There isn't a standard procedure to apply for a research position, and it can even vary within the same lab. An application might be as informal as "email a transcript, resume, and short statement that explains why you are interested in the project," or it might include an online application with several short-answer questions.

In some labs, PIs choose all undergraduate researchers, while in others the mentor who will work with the student conducts the selection process. Each PI (or lab member interested in training an undergraduate researcher) determines the selection criteria, application process, and interview procedure that works for him.

Regardless of the application procedure, it's important to remember that every part of it is a test. It's designed not only to review your qualifications but also to determine if you have the elevated level of professionalism valued in a professional research lab. Essentially, the PI's ultimate goal is to determine, "Will this student bring value to my research team, or will they bring chaos?" The take-home message is this: every part of the application procedure is an opportunity for you to excel and stand out or tank your chances of being selected for a research position.

Continue to use the hours you scheduled for a research experience in chapter 5, Your Search Strategy, to complete the application process. This is important for two reasons: (1) it will help you prioritize your applications; and (2) you'll continue to test whether or not you can commit to your proposed research hours in a real research position. All PIs prefer undergraduates who are able to manage a research commitment without compromising their academics. Saying you can manage both makes less of an impact than saying *why you know you can.* If you can explain how you've determined your available hours at an interview (from scheduling them to giving them a trial run), you'll set yourself apart from the students who can't.

Step 1: Craft Your CV and Obtain Your Transcript

As an undergraduate applying to your first research position, your accomplishments might be more appropriate for a resume than a curriculum vitae or CV. However, as you take advantage of the professional development opportunities available as part of an in-depth research experience, you'll probably find that CV will become the more accurate label. At the start, however, **it doesn't matter what you call it — CV or resume — unless you apply to a research position from an advertisement that specifically requests one or the other.** In which case, label your document whatever is instructed. In the interest in brevity, the term *CV* will be used to represent both document types for the remainder of this book.

Although there is nothing exciting about crafting your CV, the majority of PIs use one as part of their application procedure. Therefore, you need one — a good one.

It's relatively easy to craft a high-quality CV from a template provided with most word processing programs or downloaded from the Internet. Therefore, a poorly constructed one sends the message that you were either too busy (not a good sign) or too lazy (even worse) or unable (yikes!) to do so. **As tempting as it is, don't cut corners here.** Although a less-than-perfect CV probably won't prevent you from getting an interview, **a well-crafted one will give you the advantage over all students who submit a subpar one.** This is because a CV is so much more than a compilation of your accomplishments and activities. It represents you. It showcases your accomplishments, shows how seriously you take the research opportunity, and demonstrates that you believe details are important.

Things to Remember When Crafting Your CV

Before you start, keep in mind that a CV distills a vast amount of effort into a few lines of text and almost always makes its creator feel inadequate. So don't worry about what your friends puts on theirs or feel down about what you have (or don't have) to put on yours. **As you create your CV remember these points:**

- **It takes time, effort, and experiences to grow a CV.** Few students have a distinctive CV at the start of their college career. A research experience will present numerous opportunities for professional development and will help yours grow.
- **Resist the urge to misrepresent.** Unfortunately, some students feel pressured to add or inflate accomplishments, skills, or activities to make a CV more impressive. It's unnecessary and could backfire. It's better to leave it off if it isn't true, and don't exaggerate when it is.
- **Seasoned researchers understand.** Most PIs won't make their decision based on your activities, accomplishments, or skills (unless a research project has specific requirements). After all, they have "been there" and know that it's just a fact that opportunities build CVs — not wishful thinking.

Eleven Tips for Crafting a CV

Follow these tips on how to put together a stunning CV that will put you above the competition when applying for undergraduate research positions.

1. **Find a template to customize.** Most word processing programs have a CV template you can customize quickly. Alternatively, do a web search for free templates. Make one CV to use for most (if not all) research applications and use it as a foundation to build upon for the rest of your college career.

2. **Including an "Objective" section is a matter of preference, but it won't count against you if you leave it off.** However, if you want to use the same CV for all applications, you could give the wrong impression with a too-specific objective. If you state "seeking research position in a neurobiology lab," and you apply to a position in a lab that does not do neurobiology research, then the PI might pass on your application rather than risk your being disappointed with the lab's research focus.

3. **Aim for one page in length.** One page is long enough. If you decide to make it two pages, make sure your name is on the header of the second page and that you include a page number on the second page.

4. **Academic accomplishments count.** If you haven't participated in research yet, you might not have much college "stuff" to include. Consider adding a section of "Academic Accomplishments" near the top with items such as the dean's list, GPA, any fellowships or scholarships you've received, a list of lab classes you've taken, or challenging science classes you've completed.

5. **Activities and accomplishments don't have to be research-related.** No PI will discard your application because you list accomplishments unrelated to research or because you include a few activities or awards from high school. On the debate team? Volunteered at a hospital or animal shelter? Worked in retail? Won a scholarship? Add a few items. Just make sure you put items under relevant headers. One student's CV that I (DGO) reviewed listed "traveled to France" under skills. Details matter.

6. **Proofread.** Use both the spell checker and grammar checker.

7. **Use professional feedback.** For a second opinion, visit your campus career resource center or office of undergraduate research for suggestions and a professional opinion. Friends can be too shy to point out errors and are unable to offer opinions based on professional experience.

8. **Include your name in the file name.** If the PI downloads your CV, a generic label of "CV" isn't helpful when he wants to find yours quickly. When you save your CV, put your name in the file name.

9. **Save your CV as a .pdf**, not as a .doc or .docx. The only way to ensure that the PI sees what you want her to see, how you want her

to see it, is to print to, or save as, a .pdf file. Files created as .pdf will generally look the same on any operating system; whereas, .doc or .docx files may look differently depending on the operating system or version of MS Word that is installed. You don't want your CV to look poorly formatted because the PI doesn't have the latest version of MS Word. When you send your CV as a .pdf, it shows an elevated level of professionalism that few students know to do.

10. **Email a copy of your CV to yourself.** Double-check that your .pdf file looks like it should, before you send it to someone you want to impress. If it needs adjusting, it won't take long to do it. Don't waste the effort you spent crafting a quality CV only to have formatting hiccups when it arrives in a PI's in-box.

11. **Save both versions in the cloud.** Save both the .doc (or .docx) and the .pdf versions of your CV to the cloud. (Use the same app that you used to save potential research opportunities.) This way, you'll have an archive to select from as you customize future CVs for scholarships, fellowships, and recommendation letters. It takes less than three minutes to upload, and you'll be grateful for doing it more than you can imagine.

Download an Unofficial Transcript

Download, or do a screen capture, of your transcript (unofficial is fine). Even if you're a first-semester student, you'll have a transcript (it just won't list much). **Convert your downloaded file or screen capture to a .pdf and put your name in the file name (YourName Transcript.pdf). Ensure that the entire transcript is there**—sometimes information is cut off during the download or screen capture, or the personal information (your name, major, academic year) is inadvertently left off. If a PI uses a transcript as part of the application process, he wants a complete transcript including all classes, GPA, personal information, and academic year to evaluate the complete academic picture of the applicant. Even though it may seem like a small detail, for some PIs, sending an incomplete transcript means an incomplete application.

Step 2: First Contact

Now that you have a CV, have downloaded your transcript, and have identified ten to fifteen interesting projects or research labs, it's time for First Contact. **First Contact is the first time you contact a potential PI to**

ask for a research position. First Contact can be done in person, through email, or by submitting an online application to a database. Why does First Contact matter? Every PI, graduate student, and professional researcher is busy, and most don't have time to interview all students who inquire about a research opportunity. **Therefore, in the few seconds to short minute that a PI considers your request for a research position, she evaluates four things:**

1. Where does this student rank on the "professional scale"? (decided by the first impression and overall impression)
2. Does this student demonstrate a genuine interest in what I do?
3. Does this student demonstrate the ability to learn on their own?
4. Does this student have the qualifications I require?

In other words, as you've heard before, you never get a second chance to make a first impression. **First Contact is the quintessential first impression. This is the first real test.** Do well, and you'll practically guarantee an interview.

The best way to contact a PI is a matter of opinion, related to your comfort level and the individuals involved. If you decide on First Contact by email, and customize the templates we've provided, you'll probably be able to send ten emails in the time it takes to drink a venti latte. If you decide on First Contact in person, it will take longer, and you'll want to first prepare by reading the tips in both this and the next chapter.

First Contact by Email

It's true that professors receive a lot of email—sometimes more than is humanly possible to read. However, it's an oversimplification to say that high email volume is the only reason a PI doesn't respond to a student inquiry about a research position. **Professional emails that demonstrate genuine interest in the PI's research are more likely to receive a response or an invitation to an interview if a PI has an available position.** Generic or unprofessional emails are more likely to receive a generic response or no response at all. To help you make the best First Contact impression through email, follow the advice in the next section. After reviewing that section, use the email templates provided to send customized emails that demonstrate your knowledge of the research program, genuine interest in it, and your elevated professionalism.

Eleven tips for professional emails

Remember, always email with distinction. The tips here (excluding the research-specific items) **apply to all emails you send to your professors.** It doesn't matter if you're applying for a research position, need clarification on lecture material, or want to schedule an appointment. **Being more professional is always better and can help set the stage for a recommendation letter down the road.**

1. **The details matter.** Each email you send represents *you*. You'll never offend by sending a polished, professional email. The same cannot be said for an unprofessional one. Remember, details in the lab will matter to your mentor and the PI and to the person who decides whether or not to interview you. If you send a professional email, you'll establish yourself as someone who also cares about the details.

2. **Use your official college or university email address.** Email filters can send non-university emails directly into the bulk, spam, or trash folder. Even if it's a pain, use your college or university email address to make sure your email arrives at its intended destination.

3. **Don't send generic emails.** Avoid using Bcc and Cc when contacting PIs about research positions. Mass emailing as many PIs as possible to get out your message, "I'm searching for a research experience," can be a lot of effort for little return. For some students, this approach might result in an interview, but for many this spam-a-lot approach will be unsuccessful. Being on the receiving end of spam email is the same for a PI as it is for you. No one likes it. Typically, the main reason that the mass email approach is ineffective is that it doesn't allow you to customize an email for each research position. The result is a vague, generic email that appears to be sent to random people. It's impossible for a student to convince a PI of a genuine interest in her research program if the email is also addressed to twenty other people. Therefore, avoid using Bcc, Cc, or sending generic emails.

4. **Include a relevant email subject line.** Short is especially helpful if the PI reads email on her phone. If you apply for a research position from an advertisement, put the name of the research project in the subject line. If you apply for a research position after attending a seminar, put "Saw Your Seminar. Research Position?" in the subject line. If you apply after attending a poster session, put "Saw

Your Poster. Research Position?" in the subject line. Otherwise, use general terms such as "Undergraduate Research Opportunity."

5. **Include an email salutation.** This is not the time to take an informal or unprofessional approach as it can offend the recipient. Using "Hi," "Hello," "Hey," or a variation thereof is unprofessional. **Address the PI by name and title** as "Dr. Oppenheimer" or "Professor Oppenheimer." Most people won't be offended if you opt for "Dr." even if they don't have a PhD. However, we can practically guarantee the professor will take offense if you use "Miss," "Mr." or "Mrs." In addition, using a title shows that you sent your email to a specific PI, not a mass email. Continue to use a professional title in all email communication—even if you and the PI email back and forth several times.

6. **Get to the point.** Start by declaring your interest in a research position and why you are specifically interested in the PI's research. Keep the email body short. **The longer your email, the less of it that gets read.** This may already be familiar to you: the longer your class syllabus, the less of it you probably read. PIs receive a lot of email so they appreciate brevity. Think about the amount of time you have spent reading this book. In places, you've probably just read the bold. You're busy. You prioritize. PIs are busy. They prioritize. That includes which, and how much, of their emails they read.

   If you've participated in research at the college level, a lengthy explanation of it is unnecessary in the email body. At most, add one or two sentences to your email such as, "I have previous research experience working with ______ and ______." In addition, attach a separate document to your email that describes your previous experience. In this research statement, cover (1) the number of weeks you participated, and average hours per week; (2) a short explanation of the project and how it related to the "big picture" of research in the lab; (3) what you gained from the research experience or what your favorite part of it was; and (4) details about your specific responsibilities in the lab. When you attach your research statement to your email, do so as a .pdf file with your name in the file name (YourName.PreviousResearch.pdf).

7. **Avoid a long personal backstory.** Sometimes, in an attempt to add a personal touch to an email, a student will include a lengthy personal backstory. If the backstory adds significant length, the PI will likely

skim the email or forgo reading it. In addition to adding unneeded length, *including a backstory can be risky if it inadvertently raises concerns about self-management or other personal development.* To avoid this risk if you have an interesting backstory related to research, save it for a later date such as the interview. That will give you time to carefully consider if it will help you get the research position and if it is appropriate.

8. **Don't use flowery language.** Sometimes, in an attempt to stand out or express enthusiasm for a research position, a student will use flowery language in their email. This approach can actually backfire. You may have heard that it's important to avoid flowery language in emails, but why it's best avoided, and how to recognize it, is also important to know. There are three basic reasons to avoid it. **First, it's distracting.** Flowery language distracts from the message you want to send, which is that you want to learn what the PI has to teach, and you have the enthusiasm and qualifications to make a contribution to his research program. **Second, it's highly annoying** (to some). Some PIs find flowery language so annoying that they'll quit reading an email as soon as they see it. **And finally, you don't need it.** You don't need flowery language to prove that you're smart—your transcript will do that for you. You don't need it to prove genuine enthusiasm or the ability to learn on your own. Your Impact Statement will do that for you.

   Here's how to recognize flowery language. After you write an email, read it out loud to yourself and consider how natural it sounds. Then ask yourself this: "If I asked a professor about lecture material, would I use similar words and phrases?" If not, it might be too flowery. It can be difficult to let go of a beautifully crafted statement that you worked hard to write, but if it has flowery language, set it aside to use elsewhere because it won't help you get a research position.

   In the following examples, a straightforward Impact Statement would have been more effective in driving home the message: "I want to learn, and I want to contribute." **This flowery language, unfortunately, prevents that message from getting through.**

   - As a science major, and a future physician and research scientist, I find your particularly interesting research focus to be vital to my goals of understanding the natural world. This makes your vital research an important area of study that I am uniquely qualified to pursue, through my established work ethic and determination.

- I am a most ambitious individual with the most immense desire and passion to create a synthesis of classroom and research knowledge for my own purpose and to increase the understanding of others.
- I have a competitive work ethic and perfectionist procedure for undertaking anything that I do. Is there an immediate possibility in your lab with a research opportunity for a student of my caliber?
- After reading the research project description, suddenly I felt a profound interest and deep fascination with the subject matter.
- Research is critical to building a foundational knowledge base that would support my never-ending search for understanding the inner workings of the biological world.
- I am an adamantly organized individual who hopes to find a focus for my energies and passion, being that I am known for working excessively hard to reach my goals.
- I'm a driven, detail-oriented student who aspires to fulfill my dream of having my own research clinic, and I would be honored to belong to a research project constructed by a revered scientist such as yourself.
- The purpose of the symbiosis of a research project and an eager student, such as I, myself, am is a way to contribute to furthering the knowledge that other minds arduously begin. I can't wait to start!
- Engaging in fundamental research work from conception to conclusion sounds daunting, but understanding the cellular basis, and real-world applications of actual, real research has always been a dream of mine.

9. **Check the spelling in your email before you hit send.** You've heard it before because it matters so much. Inquiring about a "reserach opprotunity" doesn't make a positive impression. It's also enough for some PIs to decide that a student might not be detail oriented. Also, use full words, not just letters or numbers: two not 2, be not b, and use correct capitalization: I not i, and skip the emojis.

10. **Prevent unwanted font changes.** Every word in your email doesn't

have to be unique—your Impact Statement is enough customization. But you'll lose credibility if you send out an email that contains multiple fonts, because it will look as if you sent the same email, with only a slight modification, to everyone. That puts your email in the Bcc category (see previous discussion). Even if your email program doesn't show multiple fonts, the PI's might see variations. *If you cut and paste any text into your email program, cut and paste 100 percent of it to avoid a font change in the middle of an email.* Alternatively, compose your email in a word processing program and cut and paste it all at once or select all text in the email and adjust the text and font size.

11. **Include an email signature.** Few students create a professional email signature, so it's an additional way for you to stand out and show an elevated level of professionalism. An email signature is used to relay important information, which helps you keep the body of your emails short. In your email program, go to the preferences to find the signature panel and create a default signature to use in all email correspondence with professors. The essential information that you should include in your email signature is as follows:

    Full name
    Major/minors
    Academic year in college or your expected graduation date
    (optional) GPA
    University email address

The email template

Each sentence of your email should highlight your genuine interest in, or knowledge about, a research topic or how you'll bring value to the specific research program. Aim for five to six sentences (seven maximum). Use the guide that follows as you write your emails, but be flexible if the sentence sequence doesn't work for you. For example, you might want to combine the opening line and the next line or ask for a position in the middle of the email. Be specific and direct rather than try to make your thoughts fit into a template. Use these steps as a guide when crafting your First Contact email.

First sentence. Get right to the point. You're interested in joining the lab or research project, and you've learned something about the research program. You don't need to introduce yourself because you've created an email signature with that information. You don't need to include information about the classes you've taken unless the classes are a requirement for the position. You'll attach your CV and transcript, which will detail all your classes and activities, so you don't need that information in the email body.

Second sentence. Reference how you learned about the research opportunity, or became interested in the research topic. Be specific or it's not effective. "I ______." Fill in the blank with the appropriate information such as, "I attended a seminar, read a poster, spoke with [person] at a research symposium, read your research interests, or read your advertisement for an undergraduate researcher."

Third sentence. Include a customized Impact Statement. Use the three sentences you highlighted about each research opportunity to help you write this. Alternatively, if you found a research opportunity through a seminar, research symposium, or other method, use the inspirational statements you wrote down as a guide. **There are essentially two types of Impact Statements:**

Impact Statement #1 : *What you hope to learn* from working in a specific lab or working on a specific project.

Impact Statement #2 : *Why you're specifically interested* in the science the lab does.

Fourth sentence. (Optional) Mention your previous research experience, if any. One or two sentences is all you need. (See the section, Eleven tips for professional emails, tip 6, Get to the point, in this chapter for more details.)

Fifth sentence. Include your anticipated time commitment. First, include the number of hours per week you can devote to research. Second, include the length of time you'd like to participate in research.

Sixth sentence. (Optional) List specific qualifications. If you respond to an advertisement for a research position and required qualifications are listed, specifically mention them. This will show that (1) you read the ad thoroughly, and (2) you indeed have the qualifications. For example, you might write, "As required for this position, I've completed chemistry 300 and its accompanying lab."

Seventh sentence. **(Optional)** If desired, you can include times you are available to meet to go over your application. If you do this, be precise when mentioning days and times you are free.

Closing line. If you want to add an additional line before your signature, make it similar to, "I look forward to hearing from you," or "Thank you for your time," or "If you don't have an available research position this semester, would it be possible to join the lab as an observer?"

Six email templates for you to customize

Customize these templates to help you write the most powerful, professional email. **But before you hit send, examine each sentence and ask yourself, "If I'm asked about this in an interview, what will my answer be?"** For example, if you write, "I saw your ad for a research assistant, read up on the topic, and found it fascinating," you might be asked at the interview: "What did you read about the topic, and specifically what about it inspired you to apply to my lab?" Always be accurate and sincere and don't misrepresent anything when you email a PI. *Remember to attach your CV, transcript, and statement of previous research to the email before sending!*

Email Template #1. Use this template after reading a PI's research interests and incorporating Impact Statement #1: *What you hope to learn* from working in a specific lab.

Dear [Dr. X]:

I would like to be considered for a research position in your lab.

From reading your research interests, I learned that your lab conducts basic research on *Arabidopsis thaliana* and uses biochemical, microbiological, and genetic approaches to answer questions about cell function. I plan to attend graduate school, so I want to learn as many bench skills and as much about the scientific process as I can and possibly contribute enough to earn authorship on a paper. I have 10 hours per week for a research project this semester and will be available for a full-time commitment this summer. Ideally, I'd like to continue with a research project until I graduate.

I look forward to hearing from you.

Sincerely,

Emily

Emily A. Student
Major: Biochemistry. Minor: Microbiology
Class of 2019
emily.student@college.edu

Attachments: CV and transcript

Email Template #2. Use this template after reading a PI's research interests and incorporating Impact Statement #2: *Why you're specifically interested in the science the lab does.* Extra tip: Don't copy directly from the PI's research statements. Put the Impact Statement in your own words to maintain authenticity and sincerity.

Dear [Dr. X]:

From reading your research interests, I learned that you study biological electron transfer to create solar-based fuel. I would like to work on a project related to this in your lab, because I'm interested in using technology to help the environment. I have 12 to 15 hours per week available for a research commitment and plan to continue a research project until I graduate.

Do you have any spaces in your lab starting this or next semester?

Sincerely,

Michael

Michael A. Student
Major: Biochemical engineering
Class of 2019
michael.student@college.edu

Attachments: CV and transcript

Email Template #3. Use this template when answering an advertisement for a research position and incorporating Impact Statement # 2: *Why you're specifically interested in the science the lab does.* Extra tip: Shorten the Impact Statement if only the lecture course or the lab techniques are relevant. Also, some PIs advertise multiple projects, so always specifically name the project you are interested in.

Dear [Dr. X]:

I would like to apply for the advertised undergraduate research assistant position to work on the Molecular Gene Mapping project available in your lab. I'd like to participate in this project because, as I learned in my genetics course, this project will help identify the function of a gene, and I like the idea of discovering something new. Also, I enjoyed doing PCR and gel electrophoresis in my biology 201 lab, and I know that both techniques will be used extensively in the project.

I have all the qualifications listed in the advertisement: 10 hours a week for a research experience, I am self-motivated and reliable, and I am available for at least a one-year commitment. As per the instructions, I've attached a copy of my CV and unofficial transcript.

If this position has already been filled, would it be possible to join your lab as a general lab assistant or an observer until a project becomes available?

Sincerely,

Mia Student

Mia R. Student
Major: Biology
Class of 2017
mia.student@college.edu
Attachments: CV and transcript

Email Template #4. Use this template after reading the PI's research interests and recognizing techniques that you find interesting. Incorporate both Impact Statements #1 and #2: *What you hope to learn* from working in a specific lab, or working on a specific project, *and why you're specifically interested in the science the lab does.*

Dear [Dr. X]:

I would like to be considered for a research position in your lab.

When I read your research interests, I noticed that your lab uses many of the techniques that I learned in my microbiology and genetics labs. In particular, I've enjoyed working with bacterial strains, performing transformations, and doing genetic crosses with *Drosophila*. I would like more exposure to these, or similar techniques, in a research lab setting. This semester, I have 11 to 15 hours per week for a research experience, and I would like to be involved in research for 2 semesters.

Thank you for your time.

Sincerely,

Nick

Nicholas Student
Major: Microbiology. Minor: Psychology
Class of 2017
nicholas.student@college.edu

Attachments: CV and transcript

Email Template #5. Use this template after attending a seminar, and use Impact Statement #2: *Why you're specifically interested in the science the lab does.* Use one or two of the inspiring statements that you wrote down at the seminar. Extra tip: Because going to a seminar gives you an advantage, you can use the exact phrases you wrote down when the speaker was presenting without losing credibility.

Dear [Dr. X]:

I would like to be considered for an undergraduate research position in your lab.

I attended the seminar you gave on May 5 titled "Seminar Title." Although most of it was beyond my academic level, two things in particular stood out. First, [fill in topic here]. Second, I really enjoyed [fill in topic here]. If you have an available project at the undergraduate level, I would love to start this semester and continue for one year including full-time next summer. This semester, I have 11 to 15 hours per week to dedicate to research.

If you don't have an available project at the undergraduate level, would it be possible to join your lab as dishwasher or observer for the rest of this semester and start a project next semester?

Sincerely,

Anna

Anna B. Student
Majors: Biochemistry and Molecular Biology
Class of 2018
anna.student@college.edu

Attachments: CV and transcript

Email Template #6. Use this template after attending a poster session at a symposium where you did not have an on-the-spot interview. Use Impact Statement #2: *Why you're specifically interested in the science the lab does.* Use the answers you wrote down after discussing the presenter's project to write your Impact Statement. Extra tip: Because going to a symposium demonstrates effort, you don't have to translate the presenter's words into your own if you cite them as the source as shown in this email.

Dear [Dr. X]:

I attended the research symposium [Symposium Name] yesterday, and spoke with [Name of Person] about their project. [Name of Person] told me that they are trying to answer [research question], and that it's important because of [reason it's important]. I would like to work on a similar project, as an assistant or as a self-directed researcher in your lab. I have 13 to 15 hours per week to dedicate for a research experience and would like to continue with research for at least 2 semesters.

I have one semester of research experience as an assistant on a project about alternative splicing in moss. I've attached a short statement covering the details of that research experience, as well as my CV highlighting the skills I gained.

Sincerely,

Austin

Austin T. Student
Major: Genetics
Class of 2017
austin.student@college.edu

Attachments: CV and transcript and research experience

First Contact in Person

If your plan to find a research position includes asking your lecture (or lab) professor, your strategy starts on the first day of class. Do your best work and be your best self. For more details and tips return to chapter 5, Your Search Strategy, and review the Ask Your Professors section in Other Creative Ways to Identify Undergraduate Research Opportunities, and Ask Your Teaching Assistants in the same section. Also, look through the email templates in the last section. They give solid examples of Impact Statements and how to use them to your advantage. Even if you speak with a PI directly, you'll need an Impact Statement as an opening to the conversation to be the most effective and impressive.

Option 1: First Contact during office hours

Attending office hours or requesting an appointment to discuss a potential research position is a good strategy.

Three tips for attending office hours to ask for a research position

1. **Be prepared.** Bring a printed transcript and CV and read chapter 7, Your Interview Strategy, before attending office hours just in case your professor decides to conduct an interview on the spot.

2. **Timing matters.** Although it's impossible to know the best time to attend office hours, the worst times are a few days before and after an exam, or if several students are waiting to see her. Therefore, use the syllabus as a guide to avoid exam time. When you arrive, if there is a line of students waiting to speak with her, consider returning at a later date.

3. **Be direct.** Avoid starting the conversation with a fake question about lecture material. Make it clear that your priority is a research position in her lab and state your Impact Statement. "Hi Dr. X, I don't have a question about the class material, but I am interested in a research position in your lab. I know you study protein trafficking in plant cells, and I'm interested because I'd like to use live cell imaging to look at how proteins move between cellular compartments. Do you have any spaces in your lab starting this or next semester?"

After that, the ball is in your professor's court. She might conduct an interview, refer you to someone on her research team, make an appointment to discuss it later, or tell you that she isn't currently accepting new undergraduates. In any case, you'll likely either secure an interview or know to cross that lab off your list.

Option 2: Ask for an appointment

If you have a conflict and cannot attend office hours, request an appointment. However, **state that the reason for the appointment is to discuss a possible research position in his lab — don't try to "trick" him into an interview by indicating that you want to discuss class material.** It will be awkward and obvious if you ask one or two soft questions about lecture material and then bring up your research agenda. Be honest about why you want the appointment. If he doesn't have an available project, surprising him at an appointment won't make one available.

Option 3: Ask after lecture

This can be nerve racking, but it can sometimes be a quick way to determine if your professor has an available project. Just as in office hours, it's important to be prepared and be direct. **Introduce yourself, state that you're interested in a research position, and give your Impact Statement.** Be prepared, however, for your professor to pass on discussing her research program immediately after lecture.

If the professor doesn't immediately open a discussion, volunteer to email a copy of your CV and transcript even if you hand her a paper copy. She might say yes to end the conversation, or she might say yes because your Impact Statement was effective. At this point, her reasons don't matter, but the action you take does.

If she says yes to your emailing the transcript and resume, or instructs you to follow up with a member of her research team, then do it that day. Ideally no later than three hours after class ends[1], but definitely before you go to bed that night. Waiting longer doesn't speak highly of your motivation, but following up the same day shows that you're excited about the possibility and ready to get to work. Also, as soon as other students hear the professor answer yes to your inquiry, they may also contact the professor, and you want to be first or close to it.

Why your professor might not want to discuss a research position immediately after lecture. If you ask your professor in person, and she doesn't start a discussion about opportunities for undergraduates in her lab, but doesn't immediately say no, it's still a good sign. Even if she is distracted by shoving papers into her bag and quickly replies, "Yeah, send me an email," don't interpret that as negative. There are several

[1] Don't speak with your professor before lecture and then email her your CV and transcript during lecture. If she notices the email's time stamp, she'll know that you weren't doing your best work in class.

reasons professors interested in mentoring undergraduate researchers might not want to have a discussion about it at the end of lecture. **Here are six common ones:**

1. **They have a packed schedule.** PIs are busy from the moment they wake up in the morning to the moment they go to bed at night. During a typical workday, most PIs schedule more than is possible to accomplish. Therefore, most professors don't have the luxury of participating in a discussion about research opportunities right at the end of lecture even if they would like to.

2. **They might need to answer lecture questions.** After class, professors often prioritize students who have questions about lecture material before heading off to the next appointment, or before the next class needs the lecture room.

3. **It's too important of a conversation to rush,** so they prefer to schedule an appointment. For many PIs, discussing a research project, their expectations, and answering your questions is too important to discuss outside of an office visit.

4. **Someone else in the lab interviews potential undergraduate researchers.** A professor might instruct you to contact another member of their lab team, such as a graduate student or a member of the professional research staff, to discuss available research positions. If they do refer you to someone, follow up the same day through email, and *mention that you were instructed to contact them by their PI.* Copy (Cc) the PI in the email so they know you followed up right away, and their lab member knows you aren't making it up. (This happens.)

5. **They want to determine if you might be a good fit for the lab before they conduct an interview.** Most PIs want to examine a transcript (at the least) and a CV before scheduling an interview. Additionally, some PIs also require a short statement detailing why you're interested in their research program and a proposed lab schedule. It's unlikely there will be enough time at the end of lecture to go through all of this information and give it careful thought.

6. **They want to know that you're serious about research and not just looking for a recommendation letter.** Because some students ask about a PI's research program in hopes of securing a recommendation letter, many professors can't be sure if First Contact after

a lecture is genuine. True, professors love to talk about their research, but only when the person who asks *is genuinely interested*, and the conversation can be meaningful to everyone involved. PIs who teach large courses, or who have been in academia for few years, may have explained their research to many glassy eyed, yawn-stifling students who have no actual interest in listening. If the professor instructs you to make an appointment, follow up with email, or visit during office hours, it simply could be to determine if you're genuinely interested.

Option 4: Drop by a PI's lab or office unannounced, without an appointment, outside of office hours

You may have heard that your best strategy for finding a research position is to "be persistent but don't be a pest." Although this statement is true, it's impossible to know when persistence crosses into pest territory for an individual PI. Basically **if you drop by announced, and the PI is there, you'll meet one of two PIs: the one who doesn't mind, or the one who classifies spontaneous visits in the pest category.** That's not to say that the first PI will immediately interview you, and the second PI will show annoyance. It's unlikely you'll learn which category the PI fits into because you could drop by at the worst possible moment for either PI. In either case, if the PI views your spontaneous visit as annoying or highly inconvenient, you'll likely lose out on a position just by showing up, even if you might have received an interview from a First Contact email. Therefore, because you won't know how a particular PI feels about spontaneous visits, or even if the PI will be around when you drop by, go to office hours, email for an appointment, or connect with the professor after lecture. You'll never offend anyone with those approaches.

What to Do Next: Your Follow-Through Strategy

After you've completed the First Contact step, don't panic if you don't hear back right away, and don't take it personally. Few PIs will respond on the same day, and some will take a week (or more). What you do next depends on what response (or lack thereof) you receive.

- **If you are asked for additional information, follow up within twenty-four hours.** A PI will only ask for additional information from students she believes might be a good addition to her research team. Consider this a win. You've made it to the next round.
- **If you receive a "Lab is full" response from an email, respond back, thank the PI for returning your email, tell her that you're inter-**

ested in joining a project next semester (if true), and ask if you should contact her then. You might not receive a response to this email, and you'll likely find a position before the next semester, but it's a good idea to keep options open at the start of your search. If you receive no response to your follow-up email, then consider it an "absolute no" and focus your efforts on other labs.

- **If you receive no response after one week, email the PI again.** To avoid any resemblance to email shaming, which won't help you get a lab position, forward your original email, or cut and paste it and send as a new email. The PI knows if she read it and didn't respond— no need to point it out. She might simply be trying to get to it, and your gentle reminder will help her prioritize it.
- If you receive no response within six days after your follow-up email, then cross that lab off your list and pursue other opportunities.

If no one responds to your email, or everyone responds that the lab is full, do the following:

- **Review all of your Impact Statements.** Make certain that they are relevant, specific, and sincere.
- **Review the time commitment you proposed.** If you're applying to labs with highly specialized techniques, and only propose a few hours per week, or want a short-term experience, that could be the problem. You might need to pursue other types of research programs or increase the amount of time you have for research as long as it doesn't risk your GPA.
- **Review the Ten Application Mistakes to Avoid** presented earlier to make sure that nothing from that list is holding you back.
- **Review the First Contact section** and pay particular attention to the Eleven Tips for Professional Email if you are using that approach.
- **Review the section why research positions are competitive** to determine if any apply to your situation.
- **Choose a different approach** for your search from the section describing other creative ways to identify a research position.
- **Do not become discouraged.** Remember that not all of this process is under your control, and sometimes it takes a couple of rounds before you line up an interview.

Step 3: Online Applications (When Applicable)

Online applications are used by individual PIs, groups, and national and international research programs. When an online application is required, instead of thinking of it as an annoying chore, "Ugh, I have to do a bunch of work," try to think of it as a search advantage, "I know *exactly* what I need to do to be considered for this opportunity."

Perhaps the most significant advantage of an application is that it removes some of the guesswork. If there is an application, then you know that a specific research project is available. Typically, an application will also include eligibility requirements and a substantial description of the research project or program. Both will make it easy for you to determine if the opportunity or program is right for you before you apply.

Many students overlook the importance of taking the time to proofread their answers or making sure to represent their most professional self on the application. Therefore, if *you take the extra time,* those other students won't be your competition—*you'll be theirs.*

Seven Tips for Applications

1. **Don't apply for a position if you don't have the required qualifications.** If the research position has a specific requirement such as prerequisite coursework or academic standing, and you do not meet that requirement, don't apply for the position. Instead spend your time pursuing opportunities that you are eligible for. However, *if a particular qualification is preferred (but not required),* it's worth completing the application, even you don't have that qualification.

2. **Be honest about the time you're willing to commit.** If you're instructed to estimate the hours per week or number of semesters you can commit to a research project, be honest about the commitment you can make. No good will come from misrepresenting what you can or are willing to do. Even if you get an interview, it will end up being a waste of your search efforts.

3. **Prioritize finishing the application(s).** If the instructions say to send supporting materials separately from the online application, such as a CV or transcript, do it the same day you fill out the application. Some PIs only review completed applications, and if you delay sending the supporting documents, the position might be filled before you finish. Also, forgetting to do so, or waiting a few days, might move your application into the no pile, because it could

send the message that you are either too busy or not interested in the position enough to prioritize finishing the application.

4. **Send *exactly* what is requested.** For instance, if you're instructed to include a one-page CV, do not include a two-page CV. If you're instructed to send two personal references, don't send five. More is not better. Sending more than what was requested doesn't mean that you went the extra mile or are "extra qualified"— *it means that you didn't follow instructions.* Remember, any part of an application can be a test, and it often is.

5. **Give well-crafted, thoughtful answers and perform a spell check before hitting submit.** If the application includes essay questions, type them out in a word processing program, run grammar and spell check, and cut and paste back into the online form. Also double-check that each answer you give is relevant to the question asked. If a question is, "Why do you want to join this lab?" be specific. Use your Impact Statement or an expanded version of it. If a question is, "What do you hope to gain from a research experience?" give specific examples that reflect your genuine goals. If you use a story to support your point, make sure it is relevant and tie it back into the question.

6. **Double-check that every question has been answered and that all supporting documentation has been uploaded.** Some online applications won't let you click submit until all blanks are filled in and documentation is attached, so there is no possibility of turning in an incomplete application. However, many applications don't have a built-in insurance policy, so if you miss a question, or don't upload supplementary information, your application might not be considered. Unfortunately, in many cases, you'll never know that the reason was an incomplete application. It would be a shame if you weren't considered for a research position because of an avoidable mistake in the application process.

7. **Save the application questions and your answers.** Prior to submitting an online application, do a screen capture of all the information on the page. This includes the questions, your answers, and any instructions or overview prior to the application portion. This will give you an additional advantage later, if you review it just prior to an interview.

CHAPTER 7

Your Interview Strategy

Ten Interview Mistakes to Avoid

ALTHOUGH it doesn't happen often, sometimes a PI will invite an undergraduate to join the lab after First Contact. However, the vast majority of undergraduates will have a face-to-face interview.

Your interview strategy starts with knowing the mistakes others make so you can avoid them. If you make a few minor mistakes at an interview, it won't matter. However, if you make a big enough mistake, it's unlikely you'll be offered the research position. And here's the part that really stinks: you might leave the interview feeling that you nailed it, only to be shocked later when you receive the surprising rejection news. It's generally believed that an undergraduate will be the most professional, enthusiastic, and prepared when at an interview. **Therefore, if you don't make the mistakes listed next, you'll be more competitive than the student who does.**

1. **Forgetting the basics.** These are all Interview 101, but sometimes we forget the best-known bits of advice. They include the following:
    - **Don't be late.** And don't show up thirty minutes early—the ideal window is five to ten minutes early.
    - **Turn off your phone or put it on airplane mode before you arrive.** Focus is important. You don't want to be distracted

by, or tempted to glance at, a text. Plus, if your phone keeps vibrating, it will distract the interviewer.

- **Don't bring food or drink.** Including gum—no interviewer wants to watch someone work the last flavor crystal out of a stick of gum. Also, your interview might take place in a lab area where food or drink is prohibited.
- **Dress appropriately**. Dress as if you'll be working in the lab. (See the section "Select your interview outfit" later in this chapter.)
- **Wear minimal (or no) fragrance.** Skip the perfume, cologne, body spray, and scented lotion just in case the interviewer has allergies or sensitivities, or if the interview room is small.
- **Make eye contact.** You don't have to constantly stare into the interviewer's eyes, but frequent eye contact is important so she knows that you're listening. If it's difficult, focus on one eye, the bridge of her nose, or an eyebrow. The interviewer won't know the difference and might even be doing the same thing with you.
- **Don't interrupt the interviewer**. Yes, you're excited to let her know that she's describing the perfect opportunity, but always wait to talk until she has finished her statement, or you could quite literally talk yourself out of the research position. Don't interrupt to finish her sentences with your thoughts.
- **Be your best professional self**. I've interviewed numerous intelligent and enthusiastic students who were unaware that their interview style was unprofessional. **Many students believe that showing up on time, dressed up, and answering questions with enthusiasm is all that is needed to be professional. It is not.** Avoid the additional mistakes that follow to put your best professional foot forward.

2. **Oversharing.** One advantage of participating in undergraduate research is the numerous opportunities to strengthen personal development. (See Personal Development Opportunities in chapter 2 for details.) **Although no interviewer expects an undergraduate to be completely polished at the start of a research experience,** most choose students who demonstrate an elevated level of professionalism and maturity at the interview. Therefore, **avoid sharing information that might indicate a need to hone personal development—especially in the area of self-management.**

For example, I (DGO) interviewed a student who said that his greatest challenge in research would be to overcome his difficulty in working in groups, because he didn't value others' opinions. He went on to explain that he hoped to use his research experience to improve his interpersonal skills before needing them in medical school. Although it's likely that he meant to show that he was a self-aware person who understood his personal weaknesses, the message he sent was that he would be difficult to work with — *very difficult.*

Therefore, avoid statements that might raise "red flags" in the area of self-management such as these:

- "My mom calls to wake me up in the morning because I overslept for an exam last semester."
- "Honestly, my biggest challenge will be to not pout when things go wrong."
- "High school was easy because teachers were always reminding me of deadlines and telling me to get stuff done. College has been hard because now I have to do it myself, but I'm getting better at it."
- "My friend said that starting a research project helped him learn to manage his time better, so I thought I'd try it and see if it works for me because I'm really disorganized."
- "If research is like the lab classes I've taken, then I hope it will be easier to remain patient during the experiments."
- "I want to do research to make friends and have something to do in the afternoon. I don't have anything to do until my friends get back to the house."

3. **Not answering a question.** Interviewers want you to answer the question they ask. Although this sounds obvious, **many students either try to second-guess what answer the interviewer is looking for (and don't answer the question) or try to give a clever response instead of being direct**. This gives an interviewer the impression that the student has poor communication skills, is uncomfortable with the question (possibly because they misrepresented something on their CV), or isn't taking the interview seriously. **Sidestepping a question can be as much of a disadvantage as answering it in an unprofessional way**.

For example, I (DGO) often ask: "Which do you prefer—working by yourself or working in groups?" Many students answer, "I can do both!" But that doesn't answer the question I asked, which was "which do you *prefer*?" not "which can you do?" I ask this question to determine which available project would be better for the student: one that requires significant self-reliance and independence from the start, or one that involves working closely with a member of my research team each day. I want to know which the undergraduate prefers so that I can match them with the appropriate project and spend the rest of the interview discussing the relevant details.

4. **Misrepresenting yourself.** You already know why it's important to be honest on the application, but it's equally important that you don't misrepresent yourself at the interview. This is especially important when discussing your professional goals and the interviewer's expectations. In those cases, **telling the interviewer what you think she wants to hear, as opposed to the truth, or how you really feel, is a terrible strategy for finding your perfect research position.**

   For example, don't claim that you'll work twenty hours a week for the next four semesters and your dream is to be a geneticist, if you're only planning to be in the lab eight hours each week for two semesters, and you want to be a physician's assistant. A PI could have distinct training tracks for individual career paths. Therefore, misrepresenting yourself to get a position in the lab might land you on a project or training path that wouldn't be a good fit for your long-term goals. In addition, your mentor can't help you reach your goals if you aren't honest with her about what they are. And if you can't achieve your goals through the available position, you need to know.

5. **Insulting the interviewer, criticizing other people, or making value judgments.** Sounds obvious, right? Surprisingly, this is common at undergraduate research interviews. Many students haven't had much interview experience before entering college, so they are unaware of the powerful impact negative comments can make, or they are unaware that they even make any negative comments. **The most common mistakes in this category are these:**

   - **Criticizing anyone.** Whether you are talking about a lecture professor, teaching assistant, classmate, academic advisor, or a former research mentor, it's a risky move to speak negatively about anyone in an interview. **To start, even on a large campus, connections can run deep—the person you criticize might be a cousin, sibling, spouse, close friend, or valued colleague to the interviewer.** And in academia, connections extend well beyond a department or campus. A colleague of ours once interviewed a transfer student who complained about the "poor teaching" at his former college without knowing about our colleague's familial connections in the student's former department. On a personal level, criticizing a roommate, sibling, or parent is also a bad idea in a research interview. If you share a personal story, make sure it doesn't include how others have disappointed you or been a source of frustration.

- **Criticizing a previous research experience.** If your first research experience didn't work out, most interviewers will want to know the reasons. This can be tricky because you want to be honest (after all, you don't want to end up in the same type of situation that you left), but you don't want to disparage the previous lab or mentor, because it isn't professional and it won't reflect well on you. The best approach is to say something along the lines of, "It wasn't the right experience for me because ______" and fill in the blank with an appropriate reason, such as, "I want to do more benchwork" or "I need a different schedule" or "I didn't connect with the subject matter as much as I thought I would." Highlight a reason without being judgmental. Keep in mind, that the interviewer might contact your previous mentor after the interview.
- **Criticizing a major, career path, or class.** The practice of badmouthing a major, career path, or class happens regularly in interviews. If asked about your choices, answer with positive (nonjudgmental) reasons for your decisions. For instance, if you're asked why you chose biology for your major, give a short statement about why it excites you, or how it's essential for your career path. Avoid saying something such as, "I chose biology because I couldn't imagine doing something gross like microbiology." You probably won't know the full academic background of the interviewer or all the disciplines that are part of the lab's research program. Stay positive to stay on the safe side.

6. **Stating a disregard for learning.** Even the most influential mentor can't motivate a student to learn. Mentors can *inspire* a student, of course, but ultimately the motivation and discipline to learn comes from within. Thus, most interviewers try to select students who place a high personal value on learning.

   The most common ways students show a disregard for learning in research interviews are by

   - **Devaluing a class.** There are many uninteresting parts to a research experience. What gets most researchers through them is the knowledge that the boring things need to be completed to do the interesting ones. No one wants to wash lab dishes, rack pipette tips for the thousandth time, or clean out media bottles, but all are necessary. **Mentors want students who will push through the uninteresting parts of research without hesitation.** So when a student explains a hiccup on their transcript by stating a version of, "That class was really boring and a waste of my time," it's enough for some interviewers to pass on the student. It doesn't matter if it was a one-credit mandatory class, a weeder class, or one related to your major. **Never devalue a class in an interview — even if you felt it was a complete waste of time and tuition money.**
   - **Sending the message that classes are too much work.** For some research experiences, the benchwork is only the beginning. Among other tasks, a student might need to read the PI's papers, learn the theory behind techniques, plan experiments, create a poster, or write a report outside of lab time. Many PIs prefer students who want at least some of these additional opportunities as part of their research experience. Therefore, when an interviewer hears any variation of, "To get an *A* in that class, I had to do everything — go to lecture, take the quizzes, study, and read the book," **it raises the concern that the student might cap their effort in the lab — perhaps midway through the protocol — if it's too much work.** It also sends the message that the student might be more interested in listing research on their CV than investing themselves in a research experience.
   - **Asking if a research experience will make up for a classroom deficit.** You know that the relationship between the classroom and a research experience can be beneficial. However, some

students don't realize that the two complement and reinforce each other, not serve as a substitute for one another. Sometimes, an undergraduate will mention research as a way to reach improbable goals. For instance, a colleague interviewed a student who said, "I didn't do well in my cell biology class, and this is a cell biology lab, so I'm hoping research will fill in the blanks before the MCAT." **Your research experience can give you an advantage in certain classes, and supplement your classroom knowledge, but it won't serve as a substitute for it.**

7. **Arrogance.** No matter how confident you are in your research skills, even if you know you've mastered the core techniques the lab does, you absolutely must establish that you're open to learning. **A student who is arrogant in a research interview rarely receives an offer to join the lab. This is because they give the impression that they will be difficult to teach.** Even if you have more experience than any undergraduate in the history of undergraduate research, completed a research internship in high school, won an international science fair, have aced all lab classes, and participated in research elsewhere on campus, you should temper your confidence with modesty and an obvious desire to learn. This doesn't mean that you'll need to minimize your accomplishments—your CV is the perfect place to highlight those. But in the interview, you'll need to emphasize a genuine interest in the research position and indicate that your overall objective is to learn. Using the phrase, "I would love the chance to perfect those techniques" or "I want to learn more about that topic" is a good start. Once you have the position, you'll have opportunities to demonstrate your skill level and impress others by how quickly you meet the objectives of your research project.

8. **Asking the interviewer to solve a problem.** Research requires both the desire and the ability to solve problems, so avoid asking your interviewer to solve one for you. Although there are numerous examples I (PG) could share, I'll select only one. During an interview, one student asked me where the scooter parking spaces closest to the lab were located, and I suggested that he check with campus parking. He replied, "Well, I could do that, but I figured you'd want to help me." The student's lack of initiative to call the parking office or glance at a parking map indicated that he would be unlikely to reach

the level of self-reliance needed for an undergraduate researcher to excel in my lab. **Even if the problem you bring to the interview is small, you risk sending the message that solving problems or making decisions might be a personal weakness.** To be on the safe side, keep your questions relevant to the research position, the interviewer's expectations, and the research program. **(Note: If you have a serious problem, most interviewers will help you find a solution or refer you to campus resource that will help you address it.)**

9. **Ignoring the interviewer's expectations.** Interviews are exciting. A mentor's enthusiasm for her research can be contagious. You want to wrap up the search and get started on a research project. These all make it easy to overlook an expectation or hope that a requirement might be negotiable after you've joined the lab, but doing so could make you quite unhappy in the long run.

   For example, an interviewer might say that the position is a dish-washing one without the chance of "promotion" to research, or that spending every Friday night in the lab will be mandatory, or that the weekly time commitment is between sixteen and twenty hours. **It's up to you to decide if any expectation or requirement is a deal breaker, an inconvenience, or no issue at all before you accept a position.** As you think it over, remember that nothing will make you more unhappy in the long term than overlooking a significant expectation or a requirement. We assure you, that particular grain of sand will not turn into a pearl.

10. **Not following an interview strategy. An interview strategy will reduce your stress level and help you get the most out of the interview.** It will also help you demonstrate professionalism and enthusiasm to the interviewer. The rest of this chapter will help you develop your interview strategy.

The Interview

As with applications, there isn't a standard interview format for undergraduate research positions. Each person interviewing selects the format that works best for him. An interviewer might be the PI, a graduate student, postdoc, or another member of the professional research staff. If you're interviewed by the PI, it could be to assist a member of his research team, to work on an independent project, or to work on a project that you propose. If you're interviewed by someone else, it will most likely be to work directly with that person.

Ultimately, each interviewer wants to determine if you're genuinely interested in the research, if you are able to make the required time commitment, if you will be a good fit for the lab, and if there are any "red flags" that could be problematic. Some interviewers will also want to determine if your schedule is compatible with the mentor's or if you have a specific skill set or other prerequisites.

Interview Styles

If a specific project is available, most interviewers will give an overview of why the project is important and how it connects with the big picture of the lab. Most will state specific requirements (*You will do A, B, and C, and you'll work with D. The time commitment will be X hours a week for X semesters.*). After that, interview styles vary. **Some interviewers will ask several questions about your experience, transcript, or extracurricular activities, and some won't ask any questions about anything.** If the available position is an observational one, formal interviews tend to be short because the bulk of the interview takes place the first few weeks in the lab. If the available position is for a general lab assistant, or a researcher with an independent project, interviews tend to be on the longer side. **Next are generalized examples of four interview styles that you may encounter.**

Interview style #1

The interviewer uses highly technical language to describe her research program and potential projects. She might explain one or two available projects then ask which you would like to work on. Alternatively, she might tell you that Jacob works on "Project A" and Madeline works on "Project B," so you should talk with both researchers to determine which is more inspiring to you.

Interview style #2

The interviewer mentions little about his research program beyond giving an overview in straightforward language. He mostly asks you questions about your CV, transcript, experiences you've had in lab classes, or what you want to accomplish through a research experience. You know it's an interview, but it feels more like a conversation you'd have with a friend, because the atmosphere is so casual.

Interview style #3

The PI talks about her program in detail. She asks you some general questions (why you want to do research, what your long-term career goals are) and looks over your resume and transcript. **If she thinks that you have potential, she assigns you several papers to read** related to her research and instructs you to follow up after doing so. If you return to discuss the papers, and do well, you'll be invited to join the lab.

Interview style #4

The interviewer goes over an available project quickly either in technical or straightforward language. **The bulk of the interview is spent discussing your schedule,** and you're offered the position without being told what the expectations are (in which case, you'll need to ask). Unless an issue with your professionalism or enthusiasm pops up, this type of interview is merely a formality, as the decision to offer you the position was made at First Contact.

Scheduling the interview

Most likely, you'll schedule research interviews through email. Regardless, keep the following in mind:

- **Respond to an interview invitation within twenty-four hours of receiving it.** Sooner is always better. You want to demonstrate enthusiasm for the position, and a longer response time does the opposite. Also, if others respond quickly, and there are a limited number of interview slots, a slow response time could cost you the opportunity.
- **Consider your commute before proposing or selecting an interview time.** Give yourself time to review your notes and still arrive at the interview ten minutes early. (There is more on reviewing your notes in the next section.)
- **If the interviewer uses an online app for scheduling, follow all instructions.** For example, by default, some apps allow the selection of multiple appointment times, but if the interviewer instructs you to choose only one, choose only one. Remember, everything could be a test.
- **Make it easy for the interviewer to schedule time with you.** If he doesn't use an online scheduling app, be absolutely clear when you are available to reduce back-and-forth emailing. Stating: "I'm available Tuesday afternoon *at 2:00 p.m.*" is more precise than: "I'm

available on Tuesday *after 2:00 p.m.*" If you're instructed to suggest multiple interview times, include one after 5:00 p.m., or one on a weekend (unless instructed otherwise). It's good to demonstrate that you understand research isn't always a nine-to-five, Monday-through-Friday activity.

- **Once scheduled, avoid changing the interview time.** Yes, sometimes conflicts come up, but make sure it's worth risking your interview. If you want to change the appointment time, you might not be offered the option to reschedule.
- **Schedule interviews close together.** If you have multiple interviews, try to schedule them within a day or two of each other. You'll want to make a decision quickly if you're offered one or multiple research positions.
- **Give yourself plenty of time for the interview.** Try to schedule your interview with enough time to accommodate multiple possibilities.

Ideally, the interviewer will provide a time estimate when it's scheduled. However, if he doesn't, avoid asking for one. Although some interviewers will interpret your asking as a sign of organization, others will interpret it as a sign of an already overcommitted student trying to squeeze in an interview. To cover all of your bases, try to schedule the interview after your classes and obligations are wrapped up for the day. You don't want to be nervously looking at the clock or wondering if you'll be late to your next appointment if the interview runs long. If you need to cut the interviewer off to leave, or broadcast that you need him to wrap up it up for any reason, you'll send the message that you have somewhere more important to be. This will generally be interpreted that you're either overcommitted or uninterested in the research opportunity.

In addition, if you don't have anything scheduled directly after the interview, and you're offered the position, you might have the opportunity to get started in the lab that day. Some PIs like to walk a new student from their office into the lab to make introductions to the other lab members. If your mentor is in the lab, you might have the opportunity to ask a few questions about the project. You can ask what you can do to prepare before your first day, and you may be able to observe a technique or do a procedure at the research bench. At the very least, even if you're only introduced to your mentor, it breaks the ice. About half of the students I (DGO) offer a research position to start in the lab the same semester they interview. Many start directly after the interview.

How to Get the Most Out of Your Interview

Preparation is the key to a less-stressful and more successful research interview. Interviews are about the exchange of information. Obviously, you want to highlight how you'll be an asset to the lab, but you also want to learn as much as you can about the research project and relevant areas of lab culture. **Essentially, you need to leave an interview with answers to the following questions:**

- Am I genuinely interested in the available project or research experience?
- Can I meet the required time commitment without compromising my academics?
- Do I understand the basic requirements of the position and the interviewer's expectations?
- Will I likely be able to accomplish my goals with the available research position?

Prepare for the Interview—After First Contact

Even though you won't know ahead of time which type of interviewer you'll encounter, it's easy to prepare for most possibilities with a single strategy. **As soon as you send out First Contact emails, continue to use the hours you scheduled for a research experience to prepare for interviews**. You might hear back quickly, so don't put this off.

Step 1: Reexamine your expectations and what you want from a specific research experience

Even if you applied for a research position in response to an advertisement, it's unlikely that all the mentor's expectations and requirements will be covered until the interview. Read the questions that follow and carefully consider how you feel about each in case they are asked at an interview. In addition to being prepared, this will help you identify your deal breakers (if you have any) and help you clarify your expectations for a research position. **Highlight any questions like these that are important to you, or might be an issue for you if it's required for the position.**

- What do I hope to gain from a research experience?
- Why am I interested in this particular lab or project?
- Will I be willing to wash lab dishes or join the lab as an observer if I'm not offered a project to work on right away?

- How many hours each week am I willing to commit to a research experience? What is my maximum number of hours per week?
- What, if any, adjustments to my schedule can I make if my ideal schedule won't work for a mentor?
- If interviewing before drop and add, will I be willing to change my class schedule to accommodate a mentor's training schedule? What if it means a 7:00 a.m. genetics lecture?
- Ideally, do I want a short-term (a semester) experience, a longer one (at least a year), or to continue from now until I graduate?
- If given the option, do I want to start this semester or next?
- Will I be available to continue a project in the upcoming summer? Will I be willing to increase my weekly hours during the summer? What is my summer weekly hour maximum?
- Which, if any, university holidays am I willing to spend working in the lab?
- What do I anticipate will be the most challenging part of undergraduate research?
- If registering for course credit is a requirement, or not allowed, will that create a problem for me?

Step 2: Prepare a list of questions about the position and the research project

Here's the upshot: undergraduates who arrive at a research interview with appropriate questions appear to be more prepared, professional, and interested in the opportunity than undergraduates who don't. In addition, to get the most out of your interview, you'll need to ask questions if the interviewer doesn't cover important information. **Glance through the Lab Culture** (chapter 3) **and Understanding Your Expectations** (chapter 4) **sections to help you prepare questions that are important to you. Then add them to the following list:**

Essential questions: Get answers to these before leaving the interview.

- How many hours per week are required?
- What blocks of time are required?
- How many days per week, or which days per week, are required?
- Will I coordinate my schedule with another lab member?
- Who will be my direct research supervisor?
- When is the start date?

- How many semesters is the commitment?
- Can I stay in the lab until I graduate? (Ask this question only if you *might* want this option.)
- Will I be able to write an undergraduate thesis or an undergraduate research paper? (Ask only if you *know* that you want those options.)
- Will I need to write a proposal before the start of my research experience?
- Will there be an end-of-semester report, poster, or paper required?
- Will I need training in lab safety or animal handling or will I need vaccinations before I start?

Questions about the project.

- What is the name of the project I will work on, and what are some key words associated with it?
- What overall question or problem does the project address, and why is it important?
- What techniques will I use at the start of the project?

Questions about research for credit (optional).

- Can I register for course credit the first semester of research?
- Is there a research contract required by the department?

Step 3: Reevaluate your proposed research schedule

At a research interview, it's essential to answer questions about your schedule with honesty and confidence. If you've been testing your academic/life balance with the hours you set aside for research at the start of the search chapter, the next part will easy. **Ask yourself these questions: "Did the hours I scheduled for research work for me?"** and "Is my maximum number of hours per week still the same?"

If adjustments need to be made, review the information on scheduling in Your Search Strategy (chapter 5) and revise your schedule.

Step 4: Prepare for the interviewer's questions

By following the search strategy and reading this chapter, you're already prepared for many research interview questions. However, you won't be able to prepare for all ahead of time, because interviewers will ask questions about their specific research opportunity. If you're surprised by a question say, "Wow, good question. Let me think about that." Then take a moment to gather your thoughts before answering. **However, depending**

on the interviewer's style, there might be a few other topics that come up. How to prepare for them before the interview is covered here.

Questions about your lab classes. If your interviewer teaches a section of a lab class you've taken, he might have a set of questions based on the course. He might want to know if the letter grade on your transcript reflects the knowledge you gained from the class, or how interested you were in the subject material. It's unlikely that he'll ask you about every experiment—probably just one or two at the most. Prepare to talk about which lab classes you liked and why, or to discuss an experiment or technique from your current lab class. This is especially important if you referenced a lab class as your inspiration for applying to a research project.

Questions about a technique you emphasized in First Contact. *If you mentioned a specific technique in First Contact or on the application, some interviewers will ask about it.* To be prepared, and not lose credibility, know the basics about the technique. Start with your general biology textbook (or general chemistry, microbiology, genetics, etc.) or learn the essential information from a web search. You don't have to learn every detail about the technique, but you should know something about how it is carried out, the reason it is used in research, and why it is interesting to you.

Questions about an advertised project. Advertisements are an excellent resource to help prepare before an interview. If you are fortunate to have this resource, read the advertisement thoroughly and look up every word and concept you don't know. To help demonstrate your genuine interest in the project, you'll want to understand the main objective or what question it addresses and why the question is important.

This information will help you answer the question, "What inspires you about the available project?" or "What do you think this project is about?" which are reasonable to expect in an interview, even if you covered them in First Contact. Your answers will also demonstrate self-reliance and self-directed learning, which puts you at an advantage—especially if other candidates cannot answer the questions.

Questions about a low grade on a transcript. First off, know this: if you sent your transcript with your First Contact email, it was reviewed before your interview was scheduled, so the interviewer knows that there is a hiccup. Even if your transcript is reviewed in front of you, don't panic and *don't offer an explanation until you are asked to do so.* In either case, the hiccup might be irrelevant to the interviewer, or only one factor considered in the overall application process. If it's important to the interviewer, she will ask you about it. Otherwise, she won't bring it up, so there is no need for you to. If you're asked to address a transcript hiccup, be honest. There is no way to know what answer will be acceptable (or unacceptable) to a particular interviewer, so you might as well go with the truth.

Reasons an interviewer might ask about a transcript hiccup:

- *She wants to know if you disliked the subject.* That could be a problem if the class material significantly overlaps with the research project, or no problem if it's irrelevant to her research program.
- *She believes it was an easy class.* That could be a problem if you pretend the subject matter was challenging when you really just blew off the class.
- *She suspects you were overcommitted* so your academics took a hit. That could be a problem if you have a history of abandoning commitments or trying to do too much at one time.

No matter the circumstances, **after you address the reason for the hiccup, immediately state: "However, I have corrected the issue that led to the hiccup by doing *this*______" (fill in the blank with your answer). It's even better if you can then add, "And I've already determined that I have X hours per week to devote to research without compromising my academics or my research responsibilities."**

Prepare for the Interview — One Day before an Interview

These are easy things to do, but don't leave them until the day of your interview. Life has a way of throwing minor complications in your way (fire drill, morning alarm doesn't go off, roommate is loud all night) that can make it hard to focus on priorities at the last minute. **Complete these items the day before your interview.**

1. **Print the following:**

    - Anything the interviewer has instructed you to bring such as proof of a financial aid award
    - A copy of the schedule you prepared during your search
    - A copy of your transcript and CV
    - A copy of the questions you want to ask the interviewer

- A list of the classes you intend to take next semester (You don't have to know which sections or the class times, just what you plan to register for.)

2. **Determine how you'll take notes.** If you plan to use pen and paper, put them in your backpack. If you plan to use an electronic device, make sure you charge it the night before.

3. **Select your interview outfit.** Even for the interview, dress as if you'll be working in the lab. If you've taken a lab class that outlined specific dress requirements, use those as a guide. In addition, select closed-toe shoes with a stable bottom. Clothing that is too loose or too tight should also be avoided.

   For positions in our lab, students have interviewed in everything from a suit and tie, a silk dress and high heels, to flip-flops and tank tops. In each case, dressing up or down too much was a disadvantage, because it prevented a student from even observing in the lab or getting started after an interview. You're a student—it's okay to dress like one. G-rated graphic tees are fine, as are jeans and running shoes. *Dress as if you'll be working in the lab, just in case you're offered the opportunity after the interview.* (Note: If you usually wear contact lenses, go to the interview in glasses. If you have long hair, tie it back beforehand.)

4. **Review the ten interview mistakes.**

   Sometime during the day before the interview (not right before bedtime), do a thorough review of the section Ten Interview Mistakes to Avoid. You'll want to focus on positivity on the day of the interview.

Prepare for the Interview—Ten Minutes before an Interview

If you review for a few minutes before the start of the interview, you'll have a more professional interview and more confidence from the start. To put your brain into "interview mode," review basic information about the position a few minutes before the interview. Start by reading your Impact Statement, the advertisement for the research position, the application questions and answers, or the email correspondence you exchanged with the interviewer. You could also go over the questions that you want to ask the interviewer. Remember, *it's okay to be nervous,* and it's actually a good sign to the interviewer, because it means that you care about the research position. As long as your nerves don't prevent you from answering questions, consider them a little bit of "healthy fear."

Good luck! If you've done what was outlined in this chapter, you're ready!

At the Interview

Take control

After you meet the interviewer, there might be an awkward pause before the interview begins. It's not crucial, but if you have the opportunity, **this could be the time to say, "I'm excited to learn more about ______" and fill in the blank with "the available project,"** "the techniques your lab does," or "your research program." This statement tells the interviewer that you're excited to be there and that you're eager to learn. It's a simple thing to do, but it's not an obvious one. As an aside, a colleague once cut an interview short after a student's first words after the introduction were, "I don't know anything about what you do; I'm just glad you emailed me back, because no one else did."

Take notes

It doesn't matter if you take notes on paper or an electronic device as long as you get the important information down. However, if you do take notes on your computer, phone, or phablet, inform the interviewer before you start. That way, she won't wonder if you've checked out during the interview or are more interested in updating your status on social media than in what she has to say. A quick, "Is it okay if I take notes on my computer?" will suffice.

There are three basic but important reasons to take notes:

1. **To help you remember the details.** Even if you accept the research position at the interview, you might not start on your research project for several weeks. Relying on your memory for something as important as the expectations of a research position is a risky strategy. **Most mentors don't plan to go over their expectations after you start on your research project, but will likely expect you to know them.**

2. **To evaluate the position.** Interviews are exciting, and it's easy to get caught up in the competition of winning the position. Your notes will help you evaluate if the position is perfect for you or be the basis of a pros and cons list if you need to decide between two positions. The notes you take on the relevant areas of lab culture will be especially helpful in this regard.

3. **For future self-assessment**. You probably won't take exams or complete graded assignments in a research lab. For many, undergraduate research is an "easy *A*" — regardless of performance. You'll want to refer to your notes on occasion to ensure you're meeting your mentor's expectations and the responsibilities of the position as they were outlined at the interview. This will not only help you get the most out of your research experience, but also help you earn the strongest recommendation letters.

Ask questions

At some point in the interview, you'll be asked a variation of, "What questions do you have?"

Not asking a question for clarification because you're intimidated, worried that you won't seem smart, or afraid the answer might not be what you want to hear is a poor interview strategy. **Being afraid to ask the question won't change the answer — it just makes you less informed.**

Even if the interviewer has answered all of your questions, you still need to ask something because it demonstrates that you're interested in the position and have been listening to her.

Regardless of how many unanswered questions you have, if you ask the right question first, you'll have the opportunity to stand out from your competition and impress the interviewer. The first question you ask depends on what information already has been given by the interviewer.

Scenario #1: The interviewer has *not* told you the required weekly hour commitment. If the time requirements haven't been addressed, your first question should be, *"So I can meet your expectations*[1]*, can you tell me how many hours per week I will be in the lab, and in what blocks of time?"* As the interviewer responds, take notes. If you're still interested in the position, and you know you can accommodate the research schedule, say, *"Sounds good. I've already worked out my schedule, so I know I can do that."* If you can't meet the time requirements, leave off the last sentence. (You'll learn how to handle this possibility later.)

Scenario #2: The interviewer *has* told you the required weekly hour commitment, and your schedule is compatible, and you're still interested in the position. Consult your notes for accuracy, then repeat back the required time commitment. Start with, *"I'd like to make sure that*

[1] The words *your expectations* are a powerful tool when used in an interview in the right context. They will make an unforgettable impact on every interviewer.

I understand your expectations," and then fill the rest in with the previously mentioned requirements. When the interviewer confirms, your next statement should be, "Sounds good. I've already worked out my schedule, so I know I can do that."

After you've demonstrated that you understand the expected time requirements, ask the rest of your questions.

What to do if you're offered the position at the interview. If you're offered the position at the interview, and it sounds like the perfect position for you, accept it right away. If you're unsure, tell the interviewer that you'll need to think about it. If you don't want the position, you can inform the interviewer right away or email her later with your decision. However, if you want the position but cannot accept it because of a deal breaker, you'll want to address it at the interview.

What to do if a deal breaker is mentioned. Sometimes, you won't know the specific requirements of a research position until you learn about them at the interview. (If you know ahead of time that you cannot meet a requirement, then you shouldn't apply for the position or accept an interview invitation.) **If the interviewer mentions a requirement that is a deal breaker for you, the ideal time to address it is *after* you've been offered the position but before you leave the interview.** (Don't bring up a deal breaker at an interview unless you are offered the position.) **However, you must do this carefully.** Don't make the mistake of trying to negotiate or asking a derivative of, "Is there any way that you could be flexible on *this* requirement?"

A better approach is to respectfully decline the offer and state the reason without asking for special consideration. This approach lessens the chance that you'll offend the interviewer, but still makes it clear that you'd like the position *if the requirement is flexible.*

Here's how you do it: You'll want to decline the offer with a very strong statement that reflects something you learned about the project, lab, or techniques in the interview. For example, say, "I am so excited about this research project because I like the idea of using PCR to determine the chromosomal location of a gene of unknown function. However, I have only twelve hours to dedicate to research per week—any more and I'll risk becoming overcommitted. If eighteen hours per week is firm, I will, unfortunately, need to decline the offer to join the lab."

At this point, the decision is in the interviewer's court, and she might be open to a solution that works for you both. However, she might say that there is no flexibility on the issue because the requirement is a deal breaker for the position. Therefore, remember that if you use this strategy, you must be prepared to walk away from the position. **Once you declare that something is a deal breaker, there is no possibility to take it back.** Be certain that your reason is something that will prevent you from keeping your academic/life balance or that will make you unhappy—not an inconvenience.

At the End of the Interview

Most interviewers won't want you to feel pressured to accept a position, so if you aren't offered the position, you'll need to make it clear if you're interested. **As the interview draws to a close, ask, "When will I know if I get the position?"**

You can also say to the interviewer, "I hope to work with you because ______." And fill in the blank with something specific you learned at the interview. Even if you had a misstep or two at the interview, this could help turn the tide in your favor. Regardless, it leaves a strong impression and will distinguish you from the other candidates who have interviews scheduled and don't include a similar statement at the end.

After the Interview: Evaluate the Opportunity

It's important to thoroughly evaluate the research opportunity while the interview is still fresh. Remember, your goal isn't simply to get a research position but to *choose the perfect position for you.* It doesn't matter if you accepted the position at the interview, or if you don't hear back from the interviewer for a while. You need to decide how you truly feel about the research position and the idea of joining the lab before the interview fades.

In addition, if the interviewer offers you the position, you'll want to respond within forty-eight hours. The fast turnaround matters because once a position is offered, the clock starts ticking. If you take too long to accept, some interviewers will presume that you're not interested, but are afraid to say so, and might offer the position to another student. Other interviewers won't be in a hurry. The trouble is, you probably won't know how long a particular interviewer will wait before offering the position to another student. **Therefore, if you evaluate the position shortly after your interview, you'll be prepared to respond to the interviewer quickly.**

Start the evaluation process by thinking about how you felt when the interview ended. Were you inspired or hoping to join the lab because you couldn't wait to get started? Alternatively, were you bored during the interview and glad when it was over so you could leave? *How you felt* *at the end of the interview is a good indication of how you feel about the opportunity overall.*

The next step is to ask yourself, "Am I genuinely interested in the techniques, project, or subject?" and **"Can I work the required hours without compromising my academics?" and** "Do I want to make the required time commitment?" and **"Do I want the position if it is offered to me?"**

What to do if you're still unsure whether you want the position

If you're still ambivalent after evaluating the position as described, try to determine why. Is the required schedule a problem or the type of work different from what you want to do? Was the interviewer intimidating? Review the relevant categories under Lab Culture (chapter 3) and Understanding Your Expectations (chapter 4) to help identify anything that might explain your ambivalence. **If your hesitation is due to an overall feeling of nervousness about starting a research experience, know that you are in good company—most students feel the same way at the start.** It would be a shame if you turned down a perfect position because the unknown was intimidating.

If your hesitation is due to a specific reason—such as the project or techniques don't sound interesting, or there is a requirement that you cannot (or don't want to) fulfill, or the type of experiential learning offered isn't what you want—**then it's best to decline the position and keep searching for something that is compatible with your goals.** If you still can't decide, keep in mind that indecision itself is often a decision. It's okay to think it over, but if you aren't sure after two days (maximum), it's probably not the right project or lab for you.

What to do if you're offered the position and it's the perfect one for you

This is easy. **Respond immediately with, "I can't wait to get started."** Include your lab schedule to ensure that you and the interviewer have the same understanding.

What to do if you receive an offer, but have another interview scheduled

This can be tricky. On the one hand, you want to explore all of your options and make a choice based on the most information. On the other hand, research positions are competitive, and you can't be sure how long an interviewer will hold the position after making an offer. (This is why, ideally, you want to schedule multiple interviews within a two-day period.) If you determined you would like to join the lab during your post-interview evaluation, you could accept the offer. Alternatively, you could email the first interviewer back, inform him of your second interview scheduled in X days, and ask if you could give him an answer after it has been completed.

What you should not do, however, is accept an offer and continue to go on other interviews, or ignore an email with an offer.

What to do if you receive two offers

Of all the tough decisions to make in life, this is a pretty good one. If both opportunities seem like the perfect position, you'll need to use other criteria to choose one. **Examine your schedule to determine if one position is more compatible with your academic/life balance. Then review the overall expectations and the answers to the questions you asked at each interview.** Use the Internet to learn more about the techniques you'll be using in each project and reread each PI's research interests. Finish with considering what you know about the lab culture, paying special attention to the training opportunities and your ultimate goals. **If both opportunities still seem equally perfect, flip a coin. It's likely that they are both perfect positions, which is why it's so hard to choose.**

What to do if you don't want the position (or you accept the position and realize before you start that it's not the right one for you)

If it's not the right lab for you, it's not the right lab for you. Although it might feel awkward, the best way to get through turning down a research position is to do it quickly and professionally.

Therefore, **on the same day you receive the offer, or on the day you realize that accepting the position was a mistake, send a short email to decline the offer**. The interviewer won't be upset with you. He wants you to find the perfect research experience as much as you do.

What happens if you aren't offered the position

Rejection can sting—especially if you were inspired by the research topic or the interviewer's enthusiasm. However, a rejection doesn't automatically mean that you weren't qualified, enthusiastic, or professional in the interview.

There are many reasons an undergraduate might not receive an offer to join a lab after an interview. For example, maybe the project was put on hold, or the mentor no longer has time to train a new student, or someone else was deemed a better fit.

Some interviewers will inform a student that they didn't get the position at the interview. This is good because it's immediate feedback, and typically the interviewer gives a reason such as a scheduling issue, over-commitment concerns, or a goal that isn't achievable through the available research position.

Other interviewers prefer to inform students through email with a variation of, "We interviewed an abundance of qualified candidates for only one position." You have no reason to doubt the sincerity of this statement—after all, you did receive an interview, so you must have been qualified on paper. PIs and mentors don't interview undergraduates for research positions for fun or because they are bored. They interview students who they believe could be an asset to their research team. **However, it might be helpful to email the interviewer and ask for specific feedback.**

Consider the feedback if any is given. You might not receive helpful feedback, even if the interviewer felt that certain "red flags" were raised during your interview. However, if an interviewer does share a specific reason that could help you strengthen your application or improve your interview skills, it's important that you consider it with an open mind. *You don't have to agree,* it is after all only his opinion, *but you should evaluate the feedback to determine if it's true* and, if so, how you can improve. Even if it's not true, it's equally important to evaluate the reason because *you'll need to change something* to make certain the next interviewer doesn't reach the same erroneous conclusion.

Reflect on the interview. Whether or not you receive feedback from an interviewer, you should reflect on each interview to determine what went well and if there were any parts that could have gone better. Actually, this is good practice after any interview, whether for an officer position in a student club, a shadowing position, or a volunteer position. You polish your professionalism not only by going on interviews, but also by reflecting on them. For this to work, be completely honest with yourself as you consider the following:

- *Did I make any of the ten interview mistakes?* For example, did the interviewer ask me the same question several times or similar questions several times? This could mean you didn't answer a question to the interviewer's satisfaction, or he was concerned that you might be misrepresenting something.
- *Did the interview end abruptly after I answered a question,* or did the interviewer seem surprised by an answer, but didn't ask me for additional information? Reflect on what was asked and the answer

you gave. That might be the key to a misspoken statement that gave the interviewer the wrong impression.

Finally, send a thank-you email even if you aren't asked to join the lab. If you don't receive an offer to join the lab, you should still email the interviewer and thank her for her time. If true, mention that you're interested if a future opportunity becomes available and mention a specific reason from the interview. (If you're not interested in the possibility of joining the lab in the future, just thank her for her time.)

Four reasons to send a thank-you email after an interview if you don't get the position

1. **It's professional.** Thanking someone for her time after an interview is a good professional habit to develop. Plus, so few students do this, it's another way to elevate your professional status.

2. **Avoid future awkwardness**. You might be in a class taught by the interviewer at a later date, and if you handle yourself with professionalism at all stages, you'll feel less awkward at office hours or in the lecture.

3. **Potential future opportunities.** If you're professional after being turned down for a research position, a sincere thank-you email can leave the door open for future opportunities. You might not have been the right candidate at that time, or for that project, but that doesn't mean you won't be the right candidate for the next project. If you are interested in joining the lab in the future, a thank-you email can help.

4. **Connections run deep.** Sometimes an interviewer will know of a colleague who is searching for an undergraduate researcher and might pass on your name. A thank-you email demonstrates elevated professionalism, which increases the chance that he'll recommend you to a colleague.

Email Templates for after the Interview

Part of building good professional habits is thanking interviewers for their time. Send an email the same day as your interview, whether or not you've been offered a position. A template for a thank-you email follows. In addition, here are several other email templates for you to customize in response to a number of different interview outcomes.

Email template to thank the interviewer for his time

Dear [Dr. Person Who Interviewed Me],

I enjoyed meeting with you and learning about your research program and the available project in your lab. I look forward to hearing back from you once your decision has been made.

Thank you for the time you spent in the interview today.

Sincerely,

Your name
Your email signature

Email template to decline an offer

Dear [Dr. Person Who Interviewed Me],

I enjoyed meeting with you and discussing the available research project in your lab. After thinking it over, I have decided that it is not the right opportunity for me.

Thank you for the time you spent in the interview.

Sincerely,

Your name
Your email signature

Email template to accept the position

Dear [Dr. Person Who Interviewed Me],

I'm excited to accept the research position in your lab and can't wait to get started. As agreed, I'll be in the lab 15 hours per week, from 1:00 to 5:00 p.m. on M,T,W, and 1:00 to 3:00 p.m. on Fridays.

Is there anything that you suggest that I read or learn to prepare for my first day?

Sincerely,

Your name
Your email signature

Email template to ask for feedback

Dear [Dr. Person Who Interviewed Me],

I enjoyed discussing your research program even though I am disappointed that I will not be joining your lab as a researcher. Do you have any advice on how I could help prepare for future interviews or strengthen my application to be more competitive for other positions?

Thank you for the time you spent in the interview.

Sincerely,

Your name
Your email signature

Email template for deal breakers (use only after you've been offered the position)

Dear [Dr. Person Who Interviewed Me],

I am so excited about this research project because I like the idea of using PCR to determine the chromosomal location of a gene of unknown function. However, I have only 12 hours to dedicate to research per week—any more and I'll risk becoming overcommitted. If 18 hours per week is firm, I will, unfortunately, need to decline the offer to join the lab.

Sincerely,

Your name
Your email signature

CHAPTER 8

Congratulations, You're In!

CONGRATULATIONS! The start of a new research position is equal parts excitement and healthy fear. Here are tips to help you get started on your adventure as an undergraduate in the lab.

Ten Tips for Starting Your Research Experience

1. **Safety first, data second.** Complete the required safety training, vaccinations, or classes in animal handling as soon as possible. If you are unsure of how to do anything, ask a labmate or wait until someone can help you. It's always better to wait than to guess. Don't risk your safety or the safety of anyone around you.

2. **Learn about the techniques.** If you know the name of a research technique you'll be doing in the lab, try to watch an online video demonstration before your first day. Your goal isn't to memorize a technique or become an expert, but to become familiar with the language and observe it once or twice. This will make it easier when you start in the lab.

3. **Look up key words.** If the interviewer mentioned key words associated with your project, familiarize yourself with them. Your goal

isn't to memorize definitions but to acquire a basic understanding of key words and concepts. When you start in the lab, you'll be overloaded with new information, so being comfortable with project terms ahead of time will be helpful.

4. **Follow your mentor's expectations.** To have the smoothest transition into the lab, review the expectations discussed in the interview before your first day. If you were instructed to read a paper, do it before your first day even if a timeline wasn't discussed. If you're required to write a research proposal but not given a deadline, try to create an outline before your first day. If you need assistance, check with your office of undergraduate research, as many offer help with writing research proposals.

5. **Be determined to get the most out of your research experience.** Show up on time, ready to work, and ready to contribute. Take notes, ask questions, listen to feedback, and embrace the chaotic nature of research. **Review Why Choose Research, chapter 2, at the end of each semester to make sure you're on track** with reaching both your professional and personal goals.

6. **Be patient with yourself.** Remember it will take time to feel as if you belong, to learn the techniques, and to understand your project. As with all new experiences, your start in the lab will have its awkward moments, but every day will be a little easier than the last.

7. **Maintain your academic/life balance.** It is only through the conscious practice of time management and prioritizing the activities that are important to you that you will achieve and maintain a solid academic/life balance. In the third week of your research experience, or sooner if needed, revisit your online scheduling app and adjust your activities as needed. **Use the approach in the Schedule and Prioritize Your Time section in chapter 5 each semester to keep your priorities in check without becoming overextended.** The small effort spent scheduling is worth the big payoff in productivity, stress reduction, and overall happiness.

8. **Be nice to everyone—not just your research mentor and the PI.** Thank your labmates when they help you and congratulate others on their successes. Avoid gossiping about your labmates and always ask, "Is this a good time to ask a question?" if you need to interrupt someone who looks busy.

9. **It's good to try, even if it's not the right lab for you.** You'll need some time in the lab before you know if you like your research experience and are satisfied with its challenges and rewards. If it turns out not to be the right research experience for you, don't let it discourage you. Instead, make a professional departure and reopen your search with the knowledge of what you'd like to be different—and the confidence to find it.

10. **Visit UndergradInTheLab.com** for tips, tricks, and strategies related to undergraduate research and register if you want to be notified when our next book comes out, which picks up where *Getting In* leaves off. Also, follow us on Instagram (@YouInTheLab), Twitter (@YouInTheLab), or like us on facebook (facebook.com/undergradinthelab). Leave a comment because we'd love to hear from you!

Index

Made in the USA
San Bernardino, CA
23 August 2019